Procedures and Applications of DNA Sequencing

Procedures and Applications of DNA Sequencing

Edited by **Tom Lee**

New York

Published by Callisto Reference,
106 Park Avenue, Suite 200,
New York, NY 10016, USA
www.callistoreference.com

Procedures and Applications of DNA Sequencing
Edited by Tom Lee

International Standard Book Number: 978-1-63239-517-7 (Hardback)

Printed in the United States of America.

Contents

Preface

This book has been a concerted effort by a group of academicians, researchers and scientists, who have contributed their research works for the realization of the book. This book has materialized in the wake of emerging advancements and innovations in this field. Therefore, the need of the hour was to compile all the required researches and disseminate the knowledge to a broad spectrum of people comprising of students, researchers and specialists of the field.

Scientists around the world have been researching to understand the details of DNA. This book presents techniques of DNA sequencing and its function in plant, mammal and medical sciences. It includes certain chapters dedicated to DNA sequencing techniques and then it focuses on diverse functions of this technology. Several renowned experts have contributed to this book. It is intended to serve as an important resource to students and experts interested in gaining more knowledge regarding DNA sequencing.

At the end of the preface, I would like to thank the authors for their brilliant chapters and the publisher for guiding us all-through the making of the book till its final stage. Also, I would like to thank my family for providing the support and encouragement throughout my academic career and research projects.

Editor

Section 1

Methods of DNA Sequencing

Hot Start 7-Deaza-dGTP Improves Sanger Dideoxy Sequencing Data of GC-Rich Targets

Sabrina Shore, Elena Hidalgo Ashrafi and Natasha Paul

TriLink BioTechnologies, Inc.

USA

1. Introduction

DNA sequencing has developed substantially over the years into a more cost-effective and accurate technique for scientific advancement in medical diagnostics, forensics, systematics, and genomics. Sanger dideoxy sequencing is currently one of the most established and popular sequencing methods (Sanger et al., 1977). Over the years, sequencing methods have become automated, faster and more specific and now allow sequencing of difficult and unknown regions of DNA. Several protocols have been developed over the years which include new dye chemistries, use of modified nucleotide analogs, use of additives in the sequencing reaction, and variations to the sequence cycling parameters (Prober et al., 1987; Kieleczawa, 2006). These modified protocols allow for sequencing through difficult regions of DNA and may be applied in the pre-sequencing PCR step or in the actual sequencing reaction itself. Despite these improvements, there continues to be DNA regions that are problematic to sequence, such as AT, GT, GC-rich regions, regions high in secondary structure, hairpins, homopolymer regions, and regions with repetitive DNA sequence (Kieleczawa, 2006; Frey et al., 2008). These challenging DNA templates often result in ambiguous sequencing data which include false stops, compressions, weak signals, and premature termination of signal. In particular, sequences high in GC content still suffer from several of these problems despite all the advancements.

Templates high in GC content have higher melting temperatures that do not allow for adequate strand separation of the DNA duplex in standard sequencing protocols. The tendency for these sequences to form complex secondary structures, such as hairpins or G-quadruplexes (Simonsson, 2001) can prevent a DNA polymerase from processively replicating an entire stretch of sequence (Weitzmann et al., 1996). The innate secondary structure of GC-rich templates and the strength of the DNA duplex can be obstacles in sequencing reactions as well as in PCR. GC-rich PCR assays are often plagued by mis-priming and inadequate amplicon yield which in turn provide poor quality DNA samples for downstream sequencing reactions (Shore & Paul, 2010). When a DNA template high in GC content is sequenced, base compressions, weak signals from high background noise, and truncated sequencing reads are typical results. Base compressions are due to secondary structure and cause abnormal migration during the electrophoretic separation step. These fragment irregularities in migration often plague the downstream sequence analysis, where the software is unable to discriminate between the fragments (Motz et al., 2000). As a result, several advancements have been developed to address the challenge of GC-rich DNA in the pre-sequencing PCR step as well as in the sequencing reaction itself.

To overcome the higher melting temperatures of GC-rich DNA sequences, sequence cycling parameters have been adjusted to use higher temperatures. Additives such as dimethyl sulfoxide (DMSO), betaine, formamide, and glycerol are often added into sequencing reactions to reduce secondary structure and to promote strand separation (Jung et al., 2002). It is well known that the use of modified analogs such as 7-deaza-dGTP (7-deaza-2'-deoxyguanosine-5'-triphosphate) and dITP (2'-deoxyinosine-5'-triphosphate) can be used in the sequencing reaction in place of dGTP or in some combination thereof to reduce secondary structure (Motz, Svante et al., 2000). 7-deaza-dGTP is a modified dGTP analog which lacks a nitrogen at the seven position of the purine ring. The absence of this nitrogen destabilizes G-quadruplex formation by preventing Hoogsteen base pairing without affecting Watson-Crick base pairing. Alternatively, dITP alters the Watson-Crick base pairing by reducing the number of hydrogen bonds from three to two in a base pair between inosine and cytosine. This reduces the strength of GC-rich duplexes by lowering the melting temperature.

Efforts have also focused on improving the PCR amplification of GC-rich targets to provide a good quality DNA template prior to the sequencing reaction. The addition of 7-deaza-dGTP in the PCR step permanently disrupts secondary structure by incorporation of this modified analog into the amplicon. The pre-sequencing PCR amplification effectively linearizes the DNA template and prepares it for sequencing. Though the optimal ratio of 7-deaza-dGTP to dGTP in a PCR reaction has been thoroughly investigated and determined, this method often requires additional components such as a Hot Start polymerase or additives to be included into the reaction (McConlogue et al., 1988).

A recently developed Hot Start technology named CleanAmp™ employs a thermolabile group on the 3' hydroxyl of a deoxynucleoside triphosphate. The presence of the protecting group blocks low temperature primer extension, which can often be a significant problem in PCR. At higher temperatures, the protecting group is released to allow for incorporation by the DNA polymerase and for more specific amplification of the intended target (Koukhareva & Lebedev, 2009). CleanAmp™ dNTPs are Hot Start versions of standard deoxynucleotide triphosphates (dATP, dCTP, dGTP, and dTTP) while CleanAmp™ 7-deaza-dGTP applies the same concept to a modified nucleoside triphosphate, 7-deaza-dGTP. Previous results have shown that a dNTP mixture containing CleanAmp™ 7-deaza-dGTP provides a significant improvement over standard 7-deaza-dGTP in the amplification of GC-rich targets in PCR assays (Shore & Paul, 2010). CleanAmp™ 7-deaza-dGTP Mix is a formulation which contains the CleanAmp™ versions of dATP, dCTP, dGTP, dTTP, and 7-deaza-dGTP where the 7-deaza-dGTP:dGTP ratio is 3:1. CleanAmp™ 7-deaza-dGTP applies a Hot Start technology to a secondary structure reducing analog that is permanently incorporated into the PCR amplicon. Preliminary sequencing results on PCR amplicons generated using the CleanAmp™ 7-deaza-dGTP Mix have shown an improvement in sequencing reads over a standard (unmodified) 7-deaza-dGTP mix. The significant improvement in data quality when CleanAmp™ 7-deaza-dGTP Mix was compared to an analogous mix containing unmodified versions of the dNTPs instigated further experimental inquiries to identify the optimal mix composition for use prior to sequencing of PCR products.

Herein we will compare the application of a CleanAmp™ 7-deaza-dGTP Mix to a standard 7-deaza-dGTP mix for the PCR amplification of GC-rich targets in preparation for Sanger dideoxy sequencing. We show that CleanAmp™ 7-deaza-dGTP Mix provides an improvement compared to the standard version of a 7-deaza-dGTP mix and provide guidance as to the best ratio of 7-deaza-dGTP to dGTP to use for optimal PCR and downstream sequencing performance. Performance categories that weigh into this decision include measures of PCR performance, such as preliminary amplicon yield and amplicon quality, and measures of sequencing performance including percent of high quality base calls, read length, and pairwise identity. The most crucial metric for determining sequence performance is the percent high quality base calls, which provides a numerical readout of fragment resolution and sequence determination. The effects of amplicon yield and amplicon quality were also assessed to determine if these parameters directly correlate with sequencing quality.

2. Methods

The use of the CleanAmp™ technology in a pre-sequencing PCR amplification step was explored by comparing analogous reactions with standard dNTPs to investigate the effect of Hot Start activation on the PCR step. The ratio of 7-deaza-dGTP to dGTP was thoroughly investigated to determine an optimal mixture that will provide robust PCR yield and accurate Sanger dideoxy sequencing results. The effect of magnesium chloride concentration on amplicon yield and subsequent sequencing results was also investigated for mixtures with high 7-deaza-dGTP ratios, which showed low amplicon yield under normal 1.5 mM magnesium chloride concentrations. Experimental data evaluates five GC-rich targets of varying GC content and amplicon length.

2.1 PCR

The following five GC-rich targets were investigated: ACE (60% GC-rich), BRAF (74%), GNAQ (79%), GNAS (84%) (Frey, Bachmann et al., 2008) and B4GN4 (78%) (Zhang et al., 2009). PCR reactions were set up in a 50 µL total volume and consisted of 1x PCR buffer (20 mM Tris pH 8.4, 50 mM KCl; Invitrogen), 1.5 mM $MgCl_2$ (Invitrogen), 2.0 U Taq DNA polymerase (Invitrogen), 0.2 µM primers (TriLink BioTechnologies), 0.2 mM d(A,C,T), X % dGTP, and Y % 7-deaza-dGTP (TriLink Biotechnologies) where X and Y are varying percentages of 0.2 mM (X/Y = 30/70, 25/75, 20/80, 10/90, 0/100). Nucleotide mixes consisted of all standard dNTPs (New England Biolabs) or all CleanAmp™ dNTPs (TriLink BioTechnologies) and 10 ng Human Genomic DNA (Promega) as template. For PCR with the 90 and 100% 7-deaza-dGTP nucleotide mixtures, different concentrations of magnesium chloride were used: 1.5, 2, or 2.5 mM. Five replicates for each condition were prepared, each in a single, thin-walled 200-µL tube and placed in a BioRad Tetrad 2 thermal cycler with a thermal cycling protocol 95 °C (10 min); [95 °C (40 s), X °C (1 s), 72 °C (1 min)] 35x; 72 °C (7 min) where X (annealing temperature) varied according to the target (ACE 70 °C; BRAF 67 °C; B4GN4 57 °C; GNAQ 66 °C; GNAS 64 °C). The GNAQ and GNAS targets went through 40 PCR cycles. After thermal cycling, all 5 reactions were combined, and 15 µL of product was run on a 2% agarose E-gel (Invitrogen) to determine preliminary amplicon yield before PCR cleanup. The five combined PCR reactions were then cleaned up to remove excess primers and dNTPs using the Qiagen PCR Cleanup kit (Qiagen). After cleanup, 5 µL of purified samples were run on an agarose gel to visualize the final product to be sequenced. Image J software was used to integrate amplicon and off-target bands on the E-gels both before PCR cleanup and after (Abramoff, 2004).

2.2 Sanger dideoxy Sequencing

PCR products generated with 70, 75, and 80% mixtures of 7-deaza-dGTP that were submitted for sequencing were amplified using 1.5 mM $MgCl_2$. The following 90 and 100% samples which were submitted for sequencing required additional magnesium chloride in the PCR step: ACE (100% - 2 mM $MgCl_2$ for standard and CleanAmp™), B4GN4 (90% - a mixture of the 1.5 and 2 mM $MgCl_2$ for standard dNTPs and 100% - 2 mM $MgCl_2$ for standard and CleanAmp™), GNAQ (100% - 2 mM $MgCl_2$ for standard and CleanAmp™), GNAS (90 and 100% - 2 mM $MgCl_2$ for standard and CleanAmp™). Those samples not mentioned contained the standard 1.5 mM $MgCl_2$.   Each target was amplified in five individual PCR experiments and analyzed by sequencing. Samples were submitted to Eton Bioscience, Inc. (San Diego, CA) where they were quantified using a nanodrop and submitted for Sanger dideoxy sequencing with the Big Dye Terminator v3.1 (Applied Biosystems) kit used for cycle sequencing reactions. A specified amount for each PCR product was used during the sequencing reaction, depending on the number of base pairs in the amplicon: ACE (156 bp, 10 ng DNA), BRAF (185 bp, 12 ng DNA), B4GN4 (720 bp, 28 ng DNA), GNAQ (642 bp, 27 ng DNA), GNAS (242 bp, 15 ng DNA). All sequencing results were analyzed with the Genious Pro v5.5.2 sequencing software (Drummond AJ, 2011).

2.3 Data analysis

Several categories of results were examined including preliminary amplicon yield, amplicon quality, pairwise identity, read length of target, and percent of high quality base calls. All values were averaged for the five independent PCR or sequencing runs and statistical analysis was done using Graphpad Prism Software (San Diego, CA). A two tailed t-test was used to test the null hypothesis that CleanAmp™ and standard dNTPs yield the same results in each 7-deaza-dGTP mixture. A one-way ANOVA test was used to test the null hyposthesis that all 7-deaza-dGTP compositions perform equally within a group (CleanAmp™ or standard dNTPs). The Tukey-Kramer post test was also done in addition to the one-way ANOVA to compare all levels of substitution with one another (Prism). Statistical probability values are indicated on graphs and tables.

3. Results

Several experimental factors were considered to rank the performance of the PCR amplification and downstream sequencing for five different GC-rich targets with 60 to 84 % GC content. These targets were amplified using five nucleotide mixtures containing different percentages of 7-deaza-dGTP, where the nucleotide mix used in the PCR step was either unmodified (standard) or Hot Start (CleanAmp™). Experimental performance criteria sought to identify the influence of the PCR conditions on amplicon formation (yield and quality) and on Sanger sequencing data quality (percent of high quality base calls, read length, and pairwise identity) (Kieleczawa et al., 2009).

3.1 PCR specificity and yield

The first factor was assessment of the PCR yield of reactions with CleanAmp™ (or hot start dNTPs) and standard dNTPs to determine if one would provide a superior result. The graphs in Figure 1 show this comparison, where each of the five targets was PCR amplified with five different 7-deaza-dGTP blends, ranging from 70 to 100% substitution of dGTP with 7-deaza-dGTP. The preliminary amplicon yield is presented as the average relative amplicon yield for each target, where the raw data values for target band gel densitometry in each experiment were normalized to the 75% CleanAmp™ 7-deaza-dGTP mixture. Amplicon yield was quantified prior to PCR cleanup and reflects approximately how much DNA yield was generated from an independent PCR reaction. While many reactions generated sufficient amplicon for downstream sequencing from a single set-up, all reactions contained five replicates to ensure that the amplicon yield after PCR cleanup was sufficient for sequencing. It was also noted that reactions using a higher percentage of 7-deaza-dGTP often required an increase in magnesium chloride concentration for sufficient amplicon yield. Therefore, the 90 and 100% 7-deaza-dGTP brews were prepared with 1.5, 2, and 2.5 mM $MgCl_2$ to ensure that enough product was formed. From the samples with varying $MgCl_2$, the PCR product with the highest yield and least off-target per run was submitted for sequencing. Preliminary experiments indicated that altered magnesium chloride concentration in the PCR step did not significantly affect sequencing reads, so this variable was eliminated when analyzing results.

When the amplicon yield of a reaction containing standard dNTPs was compared to an analogous reaction using CleanAmp™ dNTPs for a given percentage of 7-deaza-dGTP, twelve out of twenty five reactions showed a significant improvement in amplicon yield with CleanAmp™ dNTPs (Figure 1). While there were not many statistically significant differences for the lower GC-rich targets ACE (60%) and BRAF (74%), the three highest GC-rich targets: B4GN4 (78%), GNAQ (79%), and GNAS (84%) showed an improved amplicon yield using CleanAmp™ dNTPs over standard dNTPs for several 7-deaza-dGTP compositions. In the case of B4GN4, CleanAmp™ dominates in all the 7-deaza-dGTP mixtures except 100%, where very little amplicon was formed with either type of dNTPs. Furthermore, the addition of magnesium chloride had little effect on amplicon yield for this target producing very low yields prior to PCR cleanup.

The second factor was the evaluation of several 7-deaza-dGTP nucleotide blends (from 70 to 100%) across each group (CleanAmp™ or standard dNTPs) for a given target to determine an ideal amount of 7-deaza-dGTP for optimal PCR yield. In addition to the graphs shown in Figure 1, Table 1 includes the raw data averages and numerical standard deviations for this figure. Also featured in this table are the results for statistical comparison of each of the 7-deaza-dGTP compositions with one another across each of the two groups: standard or CleanAmp™ dNTPs. The colored boxes in the tables represent which of the five 7-deaza-dGTP mixtures fared the best with each type of dNTP. The results of this analysis indicate that there was not just one composition that dominated over the others for all five targets but there were some noteworthy trends. First, it was common that the 90 or 100% compositions produced less or comparable amplicon yield than any of the lower percentage compositions despite the additional $MgCl_2$ used. Second, although several different 7-deaza-dGTP blends gave comparable amplicon yields for a given target, the blend that was optimal was not consistent from one target to the next. Results were more well-defined for the CleanAmp™ group than for standard dNTPs. For CleanAmp™ dNTPs, a 75% 7-deaza-dGTP mix produced the most

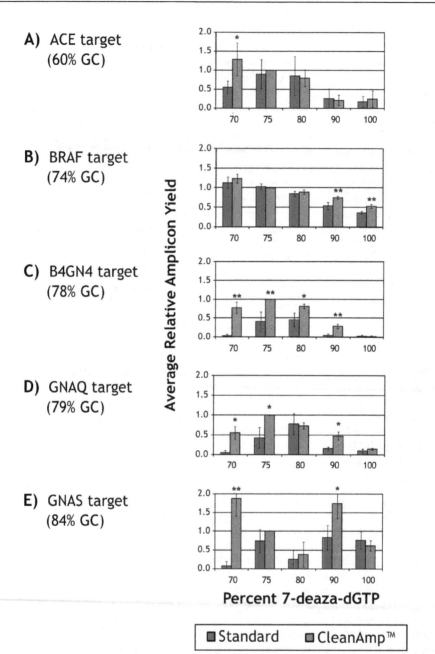

A) ACE target
(60% GC)

B) BRAF target
(74% GC)

C) B4GN4 target
(78% GC)

D) GNAQ target
(79% GC)

E) GNAS target
(84% GC)

Percent 7-deaza-dGTP

Average Relative Amplicon Yield

■ Standard   ■ CleanAmp™

Amplicon yield for each target was normalized to the 75% CleanAmp™ 7-deaza dGTP mixture, and all values were analyzed with a two tailed t-test where (* $p < 0.05$; **$p < 0.01$).

Fig. 1. Average Relative Amplicon Yield for five GC-rich targets (A-E) amplified using a dNTP mixture with 70 to 100% 7-deaza-dGTP.

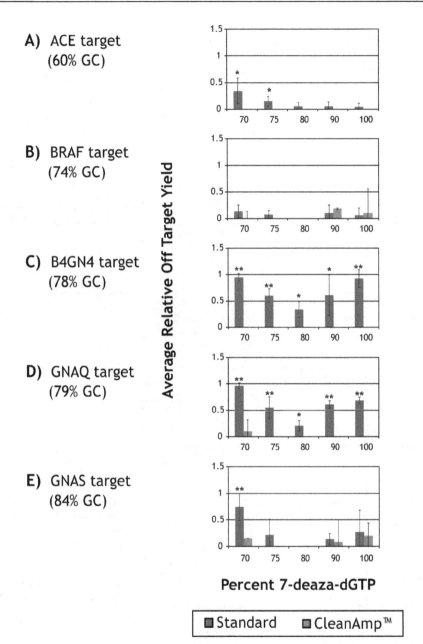

A) ACE target (60% GC)

B) BRAF target (74% GC)

C) B4GN4 target (78% GC)

D) GNAQ target (79% GC)

E) GNAS target (84% GC)

Percent 7-deaza-dGTP

■ Standard   ■ CleanAmp™

Off-Target (primer dimer and mis-priming) yield for each target was quantified relative to the desired amplicon and analyzed with a two tailed t-test where probability values (* $p < 0.05$; **$p < 0.01$) specify statistically significant values.

Fig. 2. Average Relative Off-Target Yield for five GC-rich targets (A-E) amplified using a dNTP mixture with 70 to 100% 7-deaza-dGTP.

amplicon for two targets, B4GN4 and GNAQ. A 70% 7-deaza-dGTP mix provided the highest yield for BRAF and gave higher yields than at least three other mixtures for ACE and GNAS. On the other hand, standard dNTP mixtures showed few obvious trends and varied considerably from one target to the next. For example, the B4GN4 target had comparable amplicon yields for the 75 and 80% mixtures, which provided higher yields than the remaining mixtures, while for the GNAS target, the 75, 90, and 100% yields were comparable and had greater yields than the 70 and 80% mixtures. Overall, one single 7-deaza-dGTP composition could not be identified for highest amplicon yield. In addition to amplicon yield another important factor in PCR product preparation for sequencing is amplicon quality.

Target	% 7-deaza-dGTP	Average Relative Amplicon Yield (Standard deviation)		Average Relative Off-Target (Standard deviation)	
		Standard	CleanAmp™	Standard	CleanAmp™
ACE (60% GC; 156 bp)	70	0.55 (0.17)	1.28 (0.43)	0.35 (0.24)	0 (0.0)
	75	0.90 (0.38)	1.00 (0.0)	0.15 (0.09)	0 (0.0)
	80	0.85 (0.51)	0.79 (0.22)	0.05 (0.07)	0 (0.0)
	90	0.26 (0.25)	0.21 (0.14)	0.06 (0.08)	0 (0.0)
	100	0.17 (0.15)	0.24 (0.22)	0.04 (0.07)	0 (0.0)
BRAF (74% GC; 185 bp)	70	1.13 (0.15)	1.23 (0.11)	0.14 (0.12)	0 (0.0)
	75	1.03 (0.07)	1.00 (0.0)	0.08 (0.07)	0 (0.0)
	80	0.85 (0.07)	0.89 (0.06)	0 (0.0)	0 (0.0)
	90	0.54 (0.09)	0.75 (0.04)	0.11 (0.15)	0.19 (0.42)
	100	0.37 (0.04)	0.53 (0.05)	0.06 (0.14)	0.11 (0.24)
B4GN4 (78% GC; 720 bp)	70	0.03 (0.04)	0.78 (0.15)	0.95 (0.07)	0 (0.0)
	75	0.41 (0.26)	1.00 (0.0)	0.60 (0.13)	0 (0.0)
	80	0.45 (0.18)	0.82 (0.06)	0.34 (0.16)	0 (0.0)
	90	0.04 (0.04)	0.28 (0.06)	0.61 (0.39)	0 (0.0)
	100	0.02 (0.03)	0.02 (0.02)	0.92 (0.17)	0 (0.0)
GNAQ (79% GC; 642 bp)	70	0.04 (0.06)	0.56 (0.15)	0.96 (0.05)	0.10 (0.22)
	75	0.43 (0.26)	1.00 (0.0)	0.55 (0.20)	0 (0.0)
	80	0.77 (0.26)	0.73 (0.08)	0.21 (0.10)	0 (0.0)
	90	0.15 (0.05)	0.47 (0.10)	0.61 (0.07)	0 (0.0)
	100	0.10 (0.05)	0.14 (0.02)	0.69 (0.06)	0 (0.0)
GNAS (84% GC; 242 bp)	70	0.07 (0.12)	1.87 (0.47)	0.75 (0.25)	0.15 (0.14)
	75	0.74 (0.30)	1.00 (0.0)	0.22 (0.30)	0 (0.0)
	80	0.25 (0.24)	0.38 (0.33)	0 (0.0)	0 (0.0)
	90	0.83 (0.33)	1.74 (0.40)	0.14 (0.11)	0.09 (0.02)
	100	0.76 (0.23)	0.62 (0.14)	0.27 (0.41)	0.20 (0.45)

Amplicon yields were integrated and normalized relative to the 75% CleanAmp™ 7-deaza-dGTP mixture. Off-target amplification yields are the fraction of off-target product formed relative the desired amplicon as determined by gel densitometry. The five sets of 7-deaza-dGTP compositions in each group (standard or CleanAmp™) were analyzed by a one-way ANOVA and Tukey-Kramer post test where ($p < 0.05$; $p < 0.01$; $p < 0.001$) for a given percentage of 7-deaza-dGTP. Boxes outlined in color represent means that give statistically significant values.

Table 1. Relative Average Amplicon Yield and Off-Target Integration Data

Amplicon quality indicates the purity of the sample that is being sent for sequencing. A high quality sample should contain only the DNA target to be sequenced and be free of any contaminants, excess primers, excess dNTPs, or off-target products which might interfere with the sequencing reaction. In this study, the PCR products went through a commonly used PCR cleanup process that rids the samples of excess dNTPs and primers but does not remove off-target products that were generated during the PCR. Generation of a high quality PCR product with no off-target would eliminate the more laborious step of gel purification prior to sequencing.   Therefore, in this chapter the amplicon quality was assessed by integrating the amount of average relative off-target products in the sample after the five PCR replicates were pooled, cleaned, and concentrated. Amplicon quality is represented graphically as the fraction of off-target (mis-priming and primer dimer) generated relative to the amplicon (Figure 2, Table 1), so the lower this value, the higher the sample quality will be.

Thirteen out of twenty five reactions showed significant reduction in off-target when CleanAmp™ dNTPs were used (Figure 2). The ACE off-target products consisted entirely of primer dimer while the other four targets were prone to a combination of primer dimer and mis-priming side products (Figure 3). The two lowest percent GC-rich targets, ACE and BRAF, produced the lowest amount of off-target and highest amplicon quality, with comparable performance between CleanAmp™ and standard dNTPs. The other three target reactions formed substantially more off-target, especially when amplified with standard dNTPs. Figure 2 results show that amplicon quality is highest in most cases when CleanAmp™ mixtures are used. For the B4GN4 and GNAQ targets, reactions with standard dNTPs produced significantly more off-target products than CleanAmp™ dNTPs regardless of the percent 7-deaza-dGTP. For all five amplicons, the reactions using 75 and 80% 7-deaza-dGTP produced no primer dimer or mis-priming when CleanAmp™ was used. As was the case for amplicon yield, there was not just one composition of 7-deaza-dGTP that produced the highest quality DNA product. However it was evident that the use of CleanAmp™ dNTPs reduced the amount of off-target compared to standard dNTPs, providing an amplicon with a higher chance of successfully being sequenced.

Although it is important to generate a high quality PCR product with adequate yield, other measures of the sequencing results, such as read length, pairwise identity, and percent high quality base scores should also be considered. Read length is the number of bases that were called for a given target. Optimally, this value should match the length of the reference sequence, provided it does not exceed the ~1000 base pair limits of the current Sanger dideoxy sequencing technology (Slater & Drouin, 1992; Kieleczawa, Adam et al., 2009). For some challenging targets shorter than these upper limits, the read lengths can often become truncated due to complex secondary structures or regions of DNA that the polymerase can not read through such as GC-rich regions. Results in Table 2 indicate that the average read lengths for the five GC-rich targets were comparable between CleanAmp™ and standard dNTPs in nearly every case. These results were not surprising since these five targets, which were chosen mainly for GC content are not long enough to accurately assess the impact on read length. Therefore the effect of GC content on read length remains yet to be determined in assays with longer amplicons. B4GN4, the longest target with 720 base pairs, varied the most in read length based on standard deviations, suggesting that this was one of the most difficult targets out of the group and just beginning to approach the length threshold where sequencing becomes more of a challenge. For these targets, no one 7-deaza-dGTP mixture fared better than the rest, as all of the average read length values were statistically comparable to one other. In addition to the read length of the sequence, it is critical to identify the correct bases within a sequence.

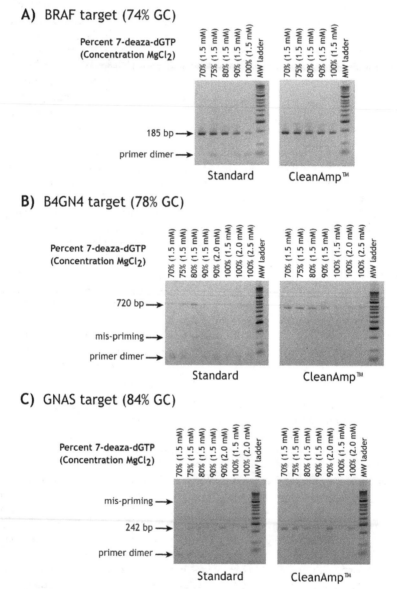

Gel images show the variability among the 3 different GC-rich targets in amplicon yield and off-target products (primer dimer and mis-priming).

Fig. 3. Agarose gel images of BRAF, B4GN4, and GNAS amplicons post PCR cleanup

3.2 Sanger dideoxy sequencing data quality

The pairwise identity in an alignment of two sequences is the percentage of shared identical bases (Drummond AJ, 2011). In this chapter, all sequences were known, so all experimental sequencing read-outs were aligned to the appropriate GenBank reference sequences (Dennis A. Benson, 2005). If the read-outs matched exactly, they would have 100% pairwise identity to the reference sequence. For ACE and BRAF, the sequencing data matched with the reference sequences nearly 100% of the time for both standard and CleanAmp™ dNTPs in all 7-deaza-dGTP compositions. For the three targets with higher GC content, results showed that standard and CleanAmp™ dNTPs pairwise identity values approached 100% with no statistically significant differences for most 7-deaza-dGTP mix compositions. However, at 70% 7-deaza-dGTP, three targets amplified with CleanAmp™ dNTPs had higher pairwise identity values than standard dNTPs. Another noteworthy outlier was B4GN4, which was prone to the highest level of off-target formation. Although no one 7-deaza-dGTP composition improved results over any of the others in the group, reactions employing CleanAmp™ dNTPs provided a higher pairwise identity to the B4GN4 reference sequences at lower 7-deaza-dGTP substitutions. While read length and correct base calls are important parameters in determining the quality of sequencing data, the most critical parameter is the percentage of high quality base calls or HQ percentage.

HQ percentage differs from pairwise identity, as it is a measure of the confidence by which the sequencing software can determine a sequence (Drummond AJ, 2011). Often, the pairwise identity may be a 100% match to the reference sequence but have very low confidence values for each base call within the sequence. The confidence in base calling becomes more important when sequencing unknown regions of DNA where there is no reference sequence. In these cases, the resultant sequence can only be trusted if the confidence of the sequencing software is high enough. Phred scores, or quality scores are a widely accepted measure of the quality of DNA sequences. Phred scores are numerical estimates of error probability for a given base (Ewing & Green, 1998). Sequencing softwares each have their own scale for base scoring, and in these studies, the percent of high quality base calls in a sequence read-out (HQ%) is defined to be the percent of bases that have a quality score (phred score) greater than 40. The highest score is a one in a million (10^{-6}) probability of a calling error where a middle score (20-40) represents a probability of only a one in a thousand (10^{-3}) (Drummond AJ, 2011). The data presented herein includes the HQ percentages for each sample (Table 2 and Figure 4).

The HQ percent scores for the ACE and BRAF sequencing data showed minimal differences between CleanAmp™ and standard dNTPs for a given 7-deaza-dGTP mix composition. For B4GN4, GNAQ, and GNAS, the sequencing of amplicons generated with CleanAmp™ dNTPs showed significantly improved HQ scores at 70% 7-deaza-dGTP relative to analogous reactions with standard dNTPs which did not reach a value of 50%. CleanAmp™ also yielded higher HQ scores ($p < 0.05$) in four out of the five mixtures for the B4GN4 target. When looking at the compositions of 7-deaza-dGTP across each group of CleanAmp™ or standard dNTPs, all HQ percent values for each 7-deaza blend are comparable for low GC content targets. For targets with greater than 75% GC content, there are some noticeable differences in the lowest (70%) and the highest (100%) substitutions of 7-deaza-dGTP. However, there were no statistically significant differences between HQ values in the middle mix compositions of 75, 80, and 90% (Table 2). Although a specific optimal percentage of 7-deaza-dGTP could not be identified, results indicate that 75, 80, and 90% blends provided the best results across a wide range of targets. Furthermore, CleanAmp™ was found to improve high quality base calls for seven out of the twenty five reactions which included certain challenging targets and lower 7-deaza-dGTP compositions for higher GC-rich species.

Target	% 7-deaza-dGTP	Average HQ% (Standard deviation)		Average Pairwise Identity (Standard deviation)		Average Read Length (Standard deviation)	
		Standard	CleanAmp™	Standard	CleanAmp™	Standard	CleanAmp™
ACE (60%)	70	69.2 (8.6)	69.5 (4.2)	99.6 (0.5)	99.8 (0.4)	108 (0)	108 (0)
	75	70.7 (6.4)	65.9 (5.4)	98.7 (1.9)	99.8 (0.4)	108 (0)	108 (0)
	80	72.0 (5.7)	67.2 (7.4)	100.0 (0)	100.0 (0)	108 (0)	108 (0)
	90	68.3 (7.0)	56.5 (25.5)	100.0 (0)	100.0 (0)	108 (0)	108 (0)
	100	56.0 (28.0)	69.9 (4.4)	99.8 (0.4)	99.8 (0.4)	108 (0)	108 (0)
BRAF (74%)	70	67.8 (7.1)	72.3 (6.3)	97.2 (2.6)	99.4 (0.7)	154.8 (2.5)	154.4 (1.1)
	75	69.7 (3.5)	78.3 (2.5)	98.2 (1.5)	99.4 (0.7)	153.8 (0.8)	153.6 (0.9)
	80	73.5 (7.6)	67.9 (7.9)	99.2 (0.8)	99.8 (0.3)	154.0 (1.4)	154.4 (0.9)
	90	67.7 (10.0)	73.6 (2.1)	98.3 (0.6)	99.3 (0.7)	154.4 (1.8)	154.0 (1.0)
	100	72.7 (5.1)	68.9 (11.5)	98.8 (1.3)	98.5 (1.4)	154.2 (1.5)	154.4 (3.4)
B4GN4 (78%)	70	18.1 (16.6)	66.3 (23.5)	55.3 (4.7)	97.1 (3.2)**	362.6 (204.1)	680.0 (2.6)*
	75	16.7 (8.9)	82.4 (6.9)	79.9 (10.9)	99.3 (0.2)	567.6 (95.4)	678.0 (1.4)
	80	29.2 (14.0)	72.6 (20.6)	86.2 (10.3)	99.0 (0.5)	667.2 (41.6)	672.8 (10.6)
	90	14.2 (12.2)	74.5 (9.9)	68.4 (10.7)	98.7 (1.2)**	535.6 (141.5)	666.4 (27.8)
	100	11.6 (6.0)	41.6 (29.0)	70.0 (16.9)	94.3 (6.6)	634.8 (158.7)	579.8 (195.4)
GNAQ (79%)	70	10.4 (19.8)	88.6 (6.0)	75.7 (17.0)	99.4 (0.7)*	508.8 (67.5)	597.6 (5.4)
	75	62.2 (35.8)	91.6 (2.6)	93.2 (10.6)	99.6 (0.2)	507.0 (204.7)	600.2 (0.5)
	80	91.3 (1.6)	92.0 (1.1)	99.5 (0.2)	99.5 (0.2)	600.0 (0)	600.8 (0.8)
	90	90.0 (3.2)	90.2 (1.1)	99.6 (0.1)	99.4 (0.1)	597.2 (5.7)	601.0 (0)
	100	77.3 (4.7)	80.4 (2.0)	99.3 (0.6)	99.5 (0.2)	600.4 (1.1)	601.0 (0.7)
GNAS (84%)	70	13.0 (19.5)	69.6 (27.1)	65.3 (17.0)	98.3 (2.8)*	162.0 (54.9)	193.2 (0.5)
	75	63.9 (35.8)	52.9 (33.2)	89.4 (20.8)	98.3 (2.8)	175.0 (25.4)	193.2 (0.5)
	80	51.6 (29.0)	49.9 (37.8)	89.6 (22.1)	90.8 (18.3)	164.4 (64.0)	195.2 (5.5)
	90	50.7 (31.2)	61.9 (8.9)	91.2 (17.4)	99.4 (0.2)	194.6 (3.6)	193.0 (0)
	100	70.7 (5.6)	79.3 (9.7)	99.0 (0.5)	99.2 (0.5)	193.2 (0.4)	188.2 (10.7)

The five sets of 7-deaza-dGTP compositions in each group (standard or CleanAmp™) were analyzed by a one-way ANOVA and Tukey-Kramer post test where probability values shown with highlighted boxes ($p < 0.05$; $p < 0.01$; $p < 0.001$) find the means statistically significant for HQ values, pairwise identity, and read length. In addition, Pairwise Identity and Read Length comparison of CleanAmp™ to standard dNTPs in individual 7-deaza-dGTP compositions, which is not shown by bar graph, are indicated by stars where (*$p < 0.05$; **$p < 0.01$; **$p < 0.001$) represent values that are statistically significant.

Table 2. Average HQ, Pairwise Identity, and Read Length Data Averages

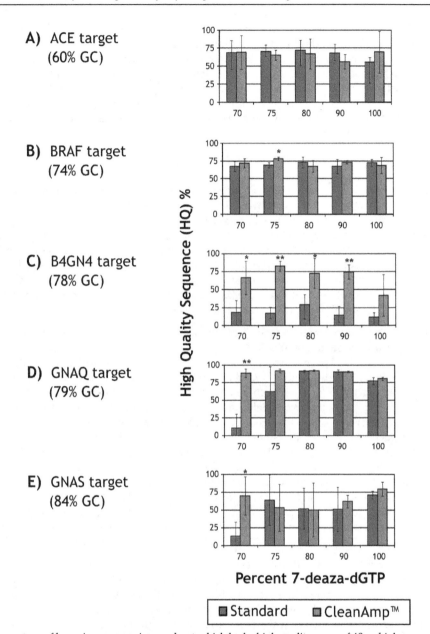

The percentage of bases in a sequencing read out which had a high quality score of 40 or higher determined a HQ percent value. These values were averaged after alignments to each reference sequence and analyzed with a two tailed t-test where probability values (* p < 0.05; **p < 0.01) find the means statistically significant.

Fig. 4. Average HQ Percentages from sequencing read-outs for all five GC-rich targets (A-E) amplified with 7-deaza-dGTP mixes of different compositions.

In summary, these studies have investigated both the percent of 7-deaza-dGTP substitution and the influence of standard and CleanAmp™ versions of the nucleotide mix on PCR and Sanger sequencing performance. When different metrics such as amplicon yield, amplicon quality, HQ percentage and sequencing chromatogram quality are considered, there are notable instances where reactions employing CleanAmp™ dNTPs have either comparable performance or statistically significant improvements in performance. To better understand how these parameters interplay with one another, a more detailed analysis will be presented in the Conclusion section.

4. Conclusion

Both standard and CleanAmp™ dNTPs can effectively generate PCR amplicons with GC-rich sequence when amplified in combination with 7-deaza-dGTP. The use of this nucleotide analog effectively linearizes the DNA sequences in preparation for Sanger dideoxy sequencing by destabilizing secondary structures such as G-quadruplexes. This allows the DNA fragments to migrate more predictably through the polyacrylamide gel and reduces the possibility of ambiguous base calls in sequencing results (Motz, Svante et al., 2000). CleanAmp™ dNTPs also offer the added benefit of reduced off-target amplification due to Hot Start activation in the PCR assay. Therefore the effect of Hot Start PCR activation in conjunction with the extent of 7-deaza-dGTP substitution was investigated to determine its potential benefit.

For the targets with lower GC content (less than 75%), CleanAmp™ dNTPs helped to reduce off-target product formation at the PCR step but gave comparable results to standard dNTPs in all other categories. For the three highest GC-rich targets, CleanAmp™ improved amplicon yield and amplicon quality with several different 7-deaza-dGTP compositions, indicating that the Hot Start activation is a much-needed benefit. In one case, B4GN4, CleanAmp™ significantly improved amplicon yield, amplicon quality, and percent HQ over standard dNTPs for at least four out of the five 7-deaza-dGTP mixtures (from 70-100% 7-deaza-dGTP). Quality sequencing results for this target were not achieved with standard dNTPs alone. The use of CleanAmp™ dNTPs at the PCR stage also improved amplicon yield, pairwise identity, and percent HQ with the 70% 7-deaza-dGTP composition. However, reactions with an analogous mixture of standard dNTPs were not as successful, indicating that standard dNTP mixtures may require a higher percentage of 7-deaza-dGTP. Although the categories of pairwise identity and read length showed minimal differences when it came to using standard or CleanAmp™ dNTPs, CleanAmp™ dNTPs improved the PCR assay and down stream sequencing results over standard dNTPs in DNA targets with GC content higher than 75%.

After analyzing the results in five categories individually, it was determined that three of the categories, amplicon yield, amplicon quality, and percent HQ, were most affected by the variables being investigated. Therefore the influence of these categories on one another was more thoroughly studied to discern the most optimal percent 7-deaza-dGTP mixture. Figure 5A(I to V) shows scatter plots of the percent HQ and amplicon yield for all variables being tested. The shaded portion of the plot highlights dNTP mixtures that reached a threshold of at least 50% relative amplicon yield and 50% high quality bases called. The dNTP compositions that were found in this region of the scatter plot were identified and re-plotted in a scatter plot of HQ percent versus amplicon quality (Figure 5B (I to V)). Optimal compositions for Figure 5B lie highest on the plots for HQ scores and furthest to the left for the least amount of off-target formed or highest amplicon quality.

Several of the different CleanAmp™ dNTP mixtures met the threshold requirements for amplicon yield, were high in amplicon quality and yielded high HQ scores. While many of the standard dNTP mixtures also have adequate amplicon yield, the amplicon quality suffered for several targets.   From the scatter plot analysis in Figure 5, the 75% CleanAmp™ mixture provided adequate amplicon yield, best amplicon quality and highest HQ scores for 4 out of the 5 targets. For the 75% mixture of CleanAmp™ dNTPs in the GNAS target adequate yield and high amplicon quality were evident but the HQ scores were lower for this composition.   The lesser correlation of GNAS to the other samples may be due to its higher GC content (84%) and the higher concentrations of magnesium chloride, which was needed for adequate amplicon yield. For targets with higher GC than 80% composition, the data indicated that complete substitution of 7-deaza-dGTP may be necessary. If more replicates are pooled to produce enough amplicon at a lower magnesium concentration then it is likely that the off-target products will decrease and base call quality will still remain high. If standard dNTPs are used in the PCR step, an 80% 7-deaza-dGTP mixture was the optimal composition. However amplicon quality can sometimes still be affected at this composition with standard dNTPs, which may require a more laborious gel purification step to remove off-target products prior to sequencing.

In addition to the numerical metrics from the sequencing data of the five targets, the sequencing chromatograms were studied. In Figure 6, representative chromatograms for the regions with high GC content for the BRAF, GNAS, and B4GN4 targets are presented. Comparisons show the optimal compositions of CleanAmp™ with 75% 7-deaza-dGTP and standard with 80% 7-deaza-dGTP. For BRAF (74% GC), a region from 50-100 bp is shown. Since the analogous HQ percentages are similar for templates with standard and CleanAmp™ dNTPs, it was not surprising that the chromatograms are similar in this region. Similarly for GNAS, the HQ percentages were comparable for both CleanAmp™ and standard dNTP samples, with a modest improvement in chromatogram shape and base call confidence for the CleanAmp™ dNTP target. For B4GN4, there were significant differences in the HQ percentages between standard (HQ: 43.7%) and CleanAmp™ (HQ: 82.2%). The sequencing trace for standard dNTPs died out at 600 bp, while the trace for CleanAmp™ dNTPs persists to the end of the target (~700 bp). Two representative regions of sequence are shown (10-60 bp and 160-210 bp), where reactions with CleanAmp™ dNTPs had strong performance for both regions, and reactions with standard dNTPs had poor performance in the early part of the read, culminating in stronger performance mid-sequence.

Overall, these studies represent a thorough investigation of both the effect of 7-deaza-dGTP substitution and the use of a Hot Start PCR technology on PCR amplification and downstream sequencing performance. Though the differences were subtle for the extent of 7-deaza-dGTP substitution when individual parameters were analyzed, the advantages of using CleanAmp™ over standard versions of the nucleotide mix, were more pronounced. Upon a more detailed analysis of amplicon yield, specificity and downstream sequencing quality, optimal nucleotide compositions were revealed. Future studies may include exploration of more targets greater than 80% in GC composition, longer templates, and the incorporation of CleanAmp™ 7-deaza-dGTP into the sequencing step.

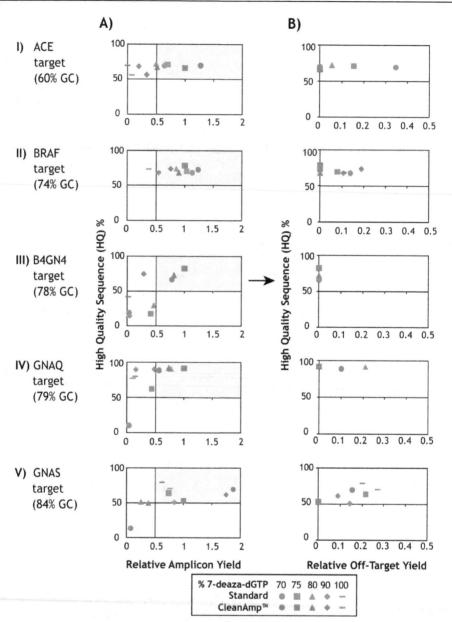

In column A, HQ and amplicon yield values that lie in the blue shaded boxes (top right) met HQ and amplicon yield threshold values and were then re-plotted with scatter plots in column B (HQ versus relative off-target yield). Values that lie furthest to the left and highest on the plots in column B are the most optimal mixtures for PCR and downstream sequencing.

Fig. 5. A) HQ-Amplicon Yield and B) HQ-relative off-target yield Scatter Plots for all five GC-rich targets (I-V).

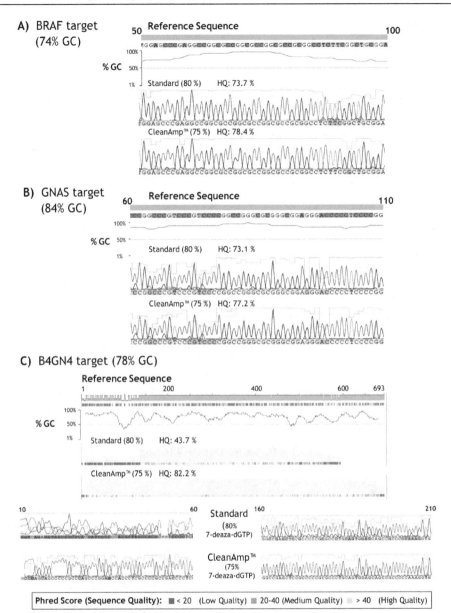

Base call quality is represented by the height of the blue shading behind the peaks which correlates with the base color (shade of blue) in the sequence. Light blue shaded bases (A, C, G, or T) are of high quality and have scores higher than 40. CleanAmp™ dNTPs show improved read length, high quality base scores, and reduced base compressions.

Fig. 6. Sequencing chromatograms of high GC-rich regions of A) BRAF, B) GNAS, and C) B4GN4 comparing standard 7-deaza-dGTP at 80% and CleanAmp™ 7-deaza-dGTP at 75% 7-deaza-dGTP substitution.

5. Acknowledgment

I would like to thank G. Zon, V. Boyd, A. McCaffrey, and R. Hogrefe for careful input and editing of this manuscript

6. References

Abramoff, M.D., Magalhaes, P.J., Ram, S.J. (2004). "Image Processing with ImageJ." *Biophotonics International* 11(7): 36-42.

Dennis A. Benson, I.K.-M., David J. Lipman, James Ostell, and David L. Wheeler (2005). "GenBank." *Nucleic Acids Res* 33(Database issue): D34-8.

Drummond AJ, A.B., Buxton S, Cheung M, Cooper A, Duran C, Field M, Heled J, Kearse M, Markowitz S, Moir R, Stones-Havas S, Sturrock S, Thierer T, Wilson A (2011). Geneious Pro v5.4, Available from http://www.geneious.com/.

Ewing, B. & Green, P. (1998). "Base-calling of automated sequencer traces using phred. II. Error probabilities." *Genome Res* 8(3): 186-94.

Frey, U.H., Bachmann, H.S., et al. (2008). "PCR-amplification of GC-rich regions: 'slowdown PCR'." *Nat Protoc* 3(8): 1312-7.

Jung, A., Ruckert, S., et al. (2002). "7-Deaza-2'-deoxyguanosine allows PCR and sequencing reactions from CpG islands." *Mol Pathol* 55(1): 55-7.

Kieleczawa, J. (2006). "Fundamentals of sequencing of difficult templates--an overview." *J Biomol Tech* 17(3): 207-17.

Kieleczawa, J., Adam, D., et al. (2009). "Identification of optimal protocols for sequencing difficult templates: results of the 2008 ABRF DNA Sequencing Research Group difficult template study 2008." *J Biomol Tech* 20(2): 116-27.

Koukhareva, I. & Lebedev, A. (2009). "3'-Protected 2'-deoxynucleoside 5'-triphosphates as a tool for heat-triggered activation of polymerase chain reaction." *Anal Chem* 81(12): 4955-62.

McConlogue, L., Brow, M.A., et al. (1988). "Structure-independent DNA amplification by PCR using 7-deaza-2'-deoxyguanosine." *Nucleic Acids Res* 16(20): 9869.

Motz, M., Svante, P., et al. (2000). "Improved Cycle Sequencing of GC-Rich Templates by a Combination of Nucleotide Analogs." *BioTechniques* 29(2): 268-270.

Prism Graphpad Prism Software. San Diego, CA USA.

Prober, J.M., Trainor, G.L., et al. (1987). "A system for rapid DNA sequencing with fluorescent chain-terminating dideoxynucleotides." *Science* 238(4825): 336-41.

Sanger, F., Nicklen, S., et al. (1977). "DNA sequencing with chain-terminating inhibitors." *Proc Natl Acad Sci U S A* 74(12): 5463-7.

Shore, S. & Paul, N. (2010). "Robust PCR amplification of GC-rich targets with Hot Start 7-deaza-dGTP." *BioTechniques* 49(5): 841-843.

Simonsson, T. (2001). "G-quadruplex DNA structures--variations on a theme." *Biol Chem* 382(4): 621-8.

Slater, G.W. & Drouin, G. (1992). "Why can we not sequence thousands of DNA bases on a polyacrylamide gel?" *Electrophoresis* 13(8): 574-82.

Weitzmann, M.N., Woodford, K.J., et al. (1996). "The development and use of a DNA polymerase arrest assay for the evaluation of parameters affecting intrastrand tetraplex formation." *J Biol Chem* 271(34): 20958-64.

Zhang, Z., Yang, X., et al. (2009). "Enhanced amplification of GC-rich DNA with two organic reagents." *BioTechniques* 47(3): 775-9.

2

DNA Representation

Bharti Rajendra Kumar

B.T. Kumaon Institute of Technology, Dwarahat,Almora, Uttarakhand,
India

1. Introduction

The term DNA sequencing refers to methods for determining the order of the nucleotides bases adenine,guanine,cytosine and thymine in a molecule of DNA. The first DNA sequence were obtained by academic researchers,using laboratories methods based on 2- dimensional chromatography in the early 1970s. By the development of dye based sequencing method with automated analysis,DNA sequencing has become easier and faster. The knowledge of DNA sequences of genes and other parts of the genome of organisms has become indispensable for basic research studying biological processes, as well as in applied fields such as diagnostic or forensic research.

DNA is the information store that ultimately dictates the structure of every gene product, delineates every part of the organisms. The order of the bases along DNA contains the complete set of instructions that make up the genetic inheritance.

The rapid speed of sequencing attained with modern DNA sequencing technology has been instrumental in the sequencing of the human genome, in the human genome project.

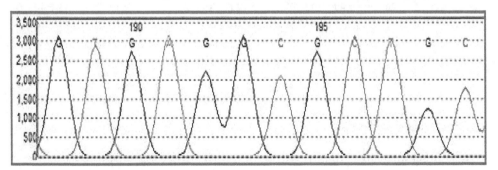

Fig. 1. DNA Sequence Trace

DNA can be sequenced by a chemical procedure that breaks a terminally labelled DNA molecule partially at each repetition of a base. The length of the labelled fragments then identify the position of that base. We describe reactions that cleave DNA preferentially at guanines,at adenines,at cytosine and thymines equally, and at cytosine alone. When the product of these four reactions are resolved by size,by electrophoresis on a polyacrylamide gel, the DNA sequences can be read from the pattern of radioactive bands. The technique will permit sequencing of atleast 100 bases from the point of labelling. The purine specific reagent is dimethyl sulphate; and the pyrimidine specific reagent is hydrazine.

In 1973 , Gilbert and Maxam reported the sequence of 24 base pairs using a method known as wandering- spot analysis.

The chain termination method developed by Sanger and coworkers in 1975 owing to its relative easy and reliability.

In 1975 the first complete DNA genome to be sequenced is that of bacteriophage φX174.

By knowing the DNA sequence, the cause of the various diseases can be known. We can determine the sequence responsible for various disease and can be treated with the help of Gene therapy.

DNA sequencing is very significant in research and forensic science. The main objective of DNA sequence generation method is to evaluate the sequencing with very high accuracy and reliability.

There are some common automated DNA sequencing problems :-

1.   Failure of the DNA sequence reaction.
2.   Mixed signal in the trace (multiple peaks).
3.   Short read lengths and poor quality data.
4.   Excessive free dye peaks "dye blobs" in the trace.
5.   Primer dimer formation in sequence reaction
6.   DNA polymerase slippage on the template mononucleotide regions.

So, we should have to do the sequencing in such a manner to avoid or minimize these problems.

DNA sequencing can solve a lot of problems and perform a lot of work for human wellfare

A sequencing can be done by different methods :

1.   Maxam – Gilbert sequencing
2.   Chain-termination methods
3.   Dye-terminator sequencing
4.   Automation and sample preperation
5.   Large scale sequencing strategies
6.   New sequencing methods.

2. Maxam-Gilbert sequencing

In 1976-1977, Allan Maxam and Walter Gilbert developed a DNA sequencing method based on chemical modification of DNA and subsequent cleavage at specific bases.

The method requires radioactive labelling at one end and purification of the DNA fragment to be sequenced. Chemical treatment generates breaks at a small proportions of one or two of the four nucleotide based in each of four reactions (G, A+G, C, C+T). Thus a series of labelled fragments is generated,from the radiolabelled end to the first 'cut' site in each molecule. The fragments   in the four reactions are arranged side by side in gel electrophoresis for size separation. To visualize the fragments,the gel is exposed to X-ray film for autoradiography,yielding a series of dark bands each corresponding to a radiolabelled DNA fragment,from which the sequence may be inferred.

3. Chain-termination method

The chain terminator method is more efficient and uses fewer toxic chemicals and lower amount of radioactivity than the method of Maxam and Gilbert.

The key principle of the Sanger method was the use of dideoxynucleotide triphosphates (ddNTPs) as DNA chain terminators.

The chain termination method requires a single-stranded DNA template,a DNA primer,a DNA polymerase, radioactively or fluorescently labelled nucleotides,and modified nucleotides that terminate DNA strand elongation. The DNA sample is divided into four separate sequencing reactions,containing all four of the standard deoxynucleotides(dATP, dGTP, dCTP, dTTP) and the DNA polymerase. To each reaction is added only one of the four dideoxynucleotide (ddATP, ddGTP, ddCTP, ddTTP) which are the chain terminating nucleotides, lacking a 3'-OH group required for the formation of a phosphodiester bond between two nucleotides,thus terminating DNA strand extension and resulting in DNA fragments of varying length.

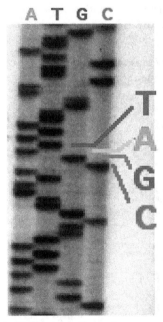

Fig. 2. Part of a radioactively labelled sequencing gel

The newly synthesized and labelled DNA fragments are heat denatured , and separated by size by gel electrophoresis on a denaturing polyacrylamide-urea gel with each of the four reactions run in one of the four individual lanes(lanes A, T, G,C), the DNA bands are then visualized by autoradiography or UV light,and the DNA sequence can be directly read off the X-ray film or gel image. A dark band in a lane indicates a DNA fragment that is result of chain termination after incorporation of a dideoxynucleotide (ddATP, ddGTP, ddCTP, or ddTTP). The relative position of the different bands among the four lanes are then used to read (from bottom to top) the DNA sequence.

The technical variations of chain termination sequencing include tagging with nucleotides containing radioactive phosphorus for labelling, or using a primer labelled at the 5′ end with a fluorescent dye. Dye- primer sequencing facilitates reading in an optical system for faster and more economical analysis and automation.

Chain termination methods have greatly simplified DNA sequencing. Limitations include non-specific binding of the primer to the DNA,affecting accurate read-out of the DNA sequence,and DNA secondary structures affecting the fidelity of the sequence.

3.1 Dye-terminator sequencing

Dye-terminator sequencing utilizes labelling of the chain terminator ddNTPs,which permits sequencing in a single reaction,rather than four reactions as in the labelled- primer method. In dye- terminator sequencing ,each of the four dideoxynucleotide chain terminators is labelled with fluorescent dyes,each of which with different wavelengths of fluorescence and emission. Owing to its greater expediency and speed,dye terminator sequencing is now the mainstay in automated sequencing. Its limitation include dye effects due to differences in the incorporation of the dye-labelled chain terminators into the DNA fragment,resulting in unequal peak heights and shapes in the electronic DNA sequence trace chromatogram.

The common challenges of DNA sequencing include poor quality in the first 15-40 bases of the sequence and deteriorating quality of sequencing traces after 700-900 bases.

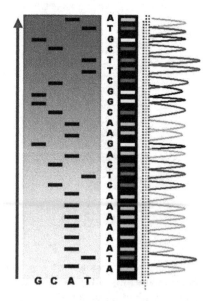

Fig. 3. Sequence ladder by radioactive sequencing compared to fluorescent peaks

3.2 Automation and sample preparation

Automated DNA sequencing instruments (DNA sequencers) can sequence upto 384 DNA samples in a single batch (run) in up to 24 runs a day. DNA sequencers carry out capillary electrophoresis for size seperation,detection and recording of dye fluorescence,and data output as fluorescent peak trace chromatograms.

A number of commercial and non-commercial software packages can trim low-quality DNA traces automatically. These programmes score the quality of each peak and remove low-quality base peaks (generally located at the ends of the sequence).

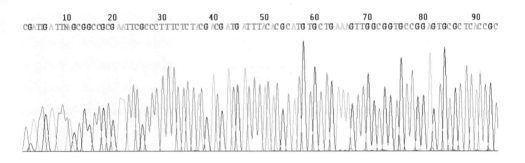

Fig. 4. View of the start of an example dye-terminator read

4. Large-scale sequencing strategies

Current methods can directly sequence only relative short (300-1000 nucleotides long) DNA fragments in a single reaction. The main obstacle to sequencing DNA fragments above this size limit is insufficient power of separation for resolving large DNA fragments that differ in length by only one nucleotide.

Large scale sequencing aims at sequencing very long DNA pieces,such as whole chromosomes. It consist of cutting (with restriction enzymes)or shearing (with mechanical forces) large DNA fragments into shorter DNA fragments. The fragmented DNA is cloned into a DNA vector, and amplified in E.coli. Short DNA fragments purified from individual bacterial colonies are individually sequenced and assembled electronically into one long,contiguous sequence. This method does not require any pre- existing information about the sequence of the DNA and is reffered to as de novo sequencing. Gaps in the assembled sequence may be filled by primer walking. The different strategies have different tradeoffs in speed and accuracy.

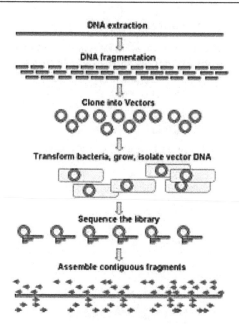

Fig. 5. Genomic DNA is fragmented into random pieces and cloned as a bacterial library. DNA from individual bacterial clones is sequenced and the sequence is assembled by using overlapping regions.

5. New sequencing methods

The high demand for low-cost sequencing has driven the development of high- throughput sequencing technologies that parallelize the sequencing process,producing thousands or millions of sequences at once. High-throughput sequencing technologies are intended to lower the cost of DNA sequencing .

Molecular detection method are not sensitive enough for single molecule sequencing, so most approaches use an in vitro cloning step to amplify individual DNA molecules.

In microfluidic Sanger sequencing the entire thermocycling amplification of DNA fragments as well as their separation by electrophoresis is done on a single chip (appoximately 100cm in diameter) thus reducing the reagent usage as well as cost. In some instances researchers have shown that they can increase the throughput of conventional sequencing through the use of microchips.

6. High throughput sequencing

The high demand for low-cost sequencing has driven the development of high-throughput sequencing technologies that parallelize the sequencing process, producing thousands or millions of sequences at once. High-throughput sequencing technologies are intended to lower the cost of DNA sequencing beyond what is possible with standard dye-terminator methods.

6.1 Lynx therapeutics' massively parallel signature sequencing (MPSS)

The first of the "next-generation" sequencing technologies, MPSS was developed in the 1990s at Lynx Therapeutics, a company founded in 1992 by Sydney Brenner and Sam Eletr. MPSS is an ultra high throughput sequencing technology. When applied to expression profile, it reveal almost every transcript in the sample and provide its accurate expression level. MPSS was a bead-based method that used a complex approach of adapter ligation followed by adapter decoding, reading the sequence in increments of four nucleotides; this method made it susceptible to sequence-specific bias or loss of specific sequences. However, the essential properties of the MPSS output were typical of later "next-gen" data types, including hundreds of thousands of short DNA sequences. In the case of MPSS, these were typically used for sequencing cDNA for measurements of gene expression levels. Lynx Therapeutics merged with Solexa in 2004, and this company was later purchased by Illumina.

6.2 Polony sequencing

It is an inexpensive but highly accurate multiplex sequencing technique that can be used to read millions of immobilized DNA sequences in parallel. This techniques was first developed by Dr. George Church in Harvard Medical college. It combined an in vitro paired-tag library with emulsion PCR, an automated microscope, and ligation-based sequencing chemistry to sequence an E. coli genome at an accuracy of > 99.9999% and a cost approximately 1/10 that of Sanger sequencing.

6.3 Pyrosequencing

A parallelized version of pyrosequencing was developed by 454 Life Sciences, which has since been acquired by Roche Diagnostics. The method amplifies DNA inside water droplets in an oil solution (emulsion PCR), with each droplet containing a single DNA template attached to a single primer-coated bead that then forms a clonal colony. The sequencing machine contains many picolitre-volume wells each containing a single bead and sequencing enzymes. Pyrosequencing uses luciferase to generate light for detection of the individual nucleotides added to the nascent DNA, and the combined data are used to generate sequence read-outs.This technology provides intermediate read length and price per base compared to Sanger sequencing on one end and Solexa and SOLiD on the other.

6.4 Illumina (Solexa) sequencing

Solexa developed a sequencing technology based on dye terminators. In this, DNA molecule are first attached to primers on a slide and amplified, this is known as bridge amplification. Unlike pyrosequencing, the DNA can only be extended one neucleotode at a time. A camera takes images of the fluorescently labeled nucleotides, then the dye along with the terminal 3' blocker is chemically removed from the DNA, allowing the next cycle.

6.5 SOLiD sequencing

The technology for sequencing used in ABISolid sequencing is oligonucleotide ligation and detection. In this, a pool of all possible oligonucleotides of fixed length are labelled according to the sequenced position. This sequencing results to the sequences of quantities and lengths comparable to illumine sequencing.

6.6 DNA nanoball sequencing

It is high throughput sequencing technology that is used to determine the entire genomic sequence of an organisms. The method uses rolling circle replication to amplify fragments of genomic DNA molecules. This DNA sequencing allows large number of DNA nanoballs to be sequenced per run and at low reagent cost compared to other next generation sequencing platforms. However, only short sequences of DNA are determined from each DNA nanoball which makes mapping the short reads to a reference genome difficult. This technology has been used for multiple genome sequencing projects and is scheduled to be used for more.

6.7 Helioscope(TM) single molecule sequencing

Helioscope sequencing uses DNA fragments with added polyA tail adapters, which are attached to the flow cell surface. The next steps involve extension-based sequencing with cyclic washes of the flow cell with fluorescently labeled nucleotides. The reads are performed by the Helioscope sequencer. The reads are short, up to 55 bases per run, but recent improvemend of the methodology allowes more accurate reads of homopolymers and RNA sequencing.

6.8 Single molecule SMRT(TM) sequencing

SMRT sequencing is based on the sequencing by synthesis approach. The DNA is synthesisd in so calles zero-mode wave-guides (ZMWs) - small well-like containers with the capturing tools located at the bottom of the well. The sequencing is performed with use of unmodified polymerase and fluorescently labelled nucleotides flowing freely in the solution. The wells are constructed in a way that only the fluorescence occurring by the bottom of the well is detected. The fluorescent label is detached from the nucleotide at its incorporation into the DNA strand, leaving an unmodified DNA strand. The SMTR technology allows detection of nucleotide modifications. This happens through the observation of polymerase kinetics. This approach allows reads of 1000 nucleotides.

6.9 Single molecule real time (RNAP) sequencing

This method is based on RNA polymerase (RNAP), which is attached to a polystyrene bead, with distal end of sequenced DNA is attached to another bead, with both beads being placed in optical traps. RNAP motion during transcription brings the beads in closer and their relative distance changes, which can then be recorded at a single nucleotide resolution. The sequence is deduced based on the four readouts with lowered concentrations of each of the four nucleotide types.

7. Other sequencing technologies

Sequencing by hybridization is a non-enzymatic method that uses a DNA microarray. A single pool of DNA whose sequence is to be determined is fluorescently labelled and hybridized to an array containing known sequences. Strong hybridization signals from a given spot on the array identifies its sequence in the DNA being sequenced. Mass spectrometry may be used to determine mass differences between DNA fragments produced in chain-termination reactions.

Some important applications of DNA sequencing are :

1. To analyse any protein structure and function we must have the knowledge of its primary structure i.e its DNA sequence.
2. With its study we can understand the function of a specific sequence and the sequence responsible for any disease.
3. With the help of comparative DNA sequence study we can detect any mutation.
4. Kinship study.
5. DNA fingerprinting.
6. By knowing the whole genome sequence, Human genome project get completed.

The main problem with sequencing is its intactness. If we perform the sequencing of same sample with different methods the result may be different so we should have to do it in such a manner that atleast 40-50% sequence must be same of similar sample.

8. Benchmarks in DNA sequencing

- 1953 Discovery of the structure of the DNA double helix.
- 1972 Development of recombinant DNA technology, which permits isolation of defined fragments of DNA; prior to this, the only accessible samples for sequencing were from bacteriophage or virus DNA.
- 1975 The first complete DNA genome to be sequenced is that of bacteriophage φX174
- 1977 Allan Maxam and Walter Gilbert publish "DNA sequencing by chemical degradation". Fred Sanger, independently, publishes "DNA sequencing by enzymatic synthesis".
- 1980 Fred Sanger and Wally Gilbert receive the Nobel Prize in Chemistry
- EMBL-bank, the first nucleotide sequence repository, is started at the European Molecular Biology Laboratory
- 1982 Genbank starts as a public repository of DNA sequences.
- Andre Marion and Sam Eletr from Hewlett Packard start Applied Biosystems in May, which comes to dominate automated sequencing.
- Akiyoshi Wada proposes automated sequencing and gets support to build robots with help from Hitachi.
- 1984 Medical Research Council scientists decipher the complete DNA sequence of the Epstein-Barr virus, 170 kb.
- 1985 Kary Mullis and colleagues develop the polymerase chain reaction, a technique to replicate small fragments of DNA
- 1986 Leroy E. Hood's laboratory at the California Institute of Technology and Smith announce the first semi-automated DNA sequencing machine.
- 1987 Applied Biosystems markets first automated sequencing machine, the model ABI 370.
- Walter Gilbert leaves the U.S. National Research Council genome panel to start Genome Corp., with the goal of sequencing and commercializing the data.
- 1990 The U.S. National Institutes of Health (NIH) begins large-scale sequencing trials on Mycoplasma capricolum, Escherichia coli, Caenorhabditis elegans, and Saccharomyces cerevisiae (at 75 cents (US)/base).

- Barry Karger (January), Lloyd Smith (August), and Norman Dovichi (September) publish on capillary electrophoresis.
- 1991 Craig Venter develops strategy to find expressed genes with ESTs (Expressed Sequence Tags).
- Uberbacher develops GRAIL, a gene-prediction program.
- 1992 Craig Venter leaves NIH to set up The Institute for Genomic Research (TIGR).
- William Haseltine heads Human Genome Sciences, to commercialize TIGR products.
- Wellcome Trust begins participation in the Human Genome Project.
- Simon et al. develop BACs (Bacterial Artificial Chromosomes) for cloning.
- First chromosome physical maps published:
- Page et al. - Y chromosome;
- Cohen et al. chromosome 21.
- Lander - complete mouse genetic map;
- Weissenbach - complete human genetic map.
- 1993 Wellcome Trust and MRC open Sanger Centre, near Cambridge, UK.
  - The GenBank database migrates from Los Alamos (DOE) to NCBI (NIH).
- 1995 Venter, Fraser and Smith publish first sequence of free-living organism, Haemophilus influenzae (genome size of 1.8 Mb).
  - Richard Mathies et al. publish on sequencing dyes (PNAS, May).
  - Michael Reeve and Carl Fuller, thermostable polymerase for sequencing.
- 1996 International HGP partners agree to release sequence data into public databases within 24 hours.
  - International consortium releases genome sequence of yeast S. cerevisiae (genome size of 12.1 Mb).
  - Yoshihide Hayashizaki's at RIKEN completes the first set of full-length mouse cDNAs.
  - ABI introduces a capillary electrophoresis system, the ABI310 sequence analyzer.
- 1997 Blattner, Plunkett et al. publish the sequence of E. coli (genome size of 5 Mb)
- 1998 Phil Green and Brent Ewing of Washington University publish "phred" for interpreting sequencer data (in use since '95).
  - Venter starts new company "Celera"; "will sequence HG in 3 yrs for $300m."
  - Applied Biosystems introduces the 3700 capillary sequencing machine.
  - Wellcome Trust doubles support for the HGP to $330 million for 1/3 of the sequencing.
  - NIH & DOE goal: "working draft" of the human genome by 2001.
  - Sulston, Waterston et al finish sequence of C. elegans (genome size of 97Mb).
- 1999 NIH moves up completion date for rough draft, to spring 2000.
  - NIH launches the mouse genome sequencing project.
  - First sequence of human chromosome 22 published.
- 2000 Celera and collaborators sequence fruit fly Drosophila melanogaster (genome size of 180Mb) - validation of Venter's shotgun method. HGP and Celera debate issues related to data release.
  - HGP consortium publishes sequence of chromosome 21.
  - HGP & Celera jointly announce working drafts of HG sequence, promise joint publication.

- Estimates for the number of genes in the human genome range from 35,000 to 120,000. International consortium completes first plant sequence, Arabidopsis thaliana(genome size of 125 Mb).
- 2001 HGP consortium publishes Human Genome Sequence draft in Nature (15 Feb).
  - Celera publishes the Human Genome sequence.
- 2005 420,000 VariantSEQr human resequencing primer sequences published on new NCBI Probe database.
- 2007 For the first time, a set of closely related species (12 Drosophilidae) are sequenced, launching the era of phylogenomics.
  - Craig Venter publishes his full diploid genome: the first human genome to be sequenced completely.
- 2008 An international consortium launches The 1000 Genomes Project, aimed to study human genetic variability.
- 2008 Leiden University Medical Center scientists decipher the first complete DNA sequence of a woman.

9. References

[1] Olsvik O, Wahlberg J, Petterson B, et al. (January 1993). "Use of automated sequencing of polymerase chain reaction-generated amplicons to identify three types of cholera toxin subunit B in Vibrio cholerae O1 strains". J. Clin. Microbiol. 31 (1): 22–5. PMC 262614.PMID 7678018.

[2] Pettersson E, Lundeberg J, Ahmadian A (February 2009). "Generations of sequencing technologies". Genomics 93 (2): 105–11. doi:10.1016/j.ygeno.2008.10.003. PMID 18992322.

[3] Min Jou W, Haegeman G, Ysebaert M, Fiers W (May 1972). "Nucleotide sequence of the gene coding for the bacteriophage MS2 coat protein". Nature 237 (5350): 82–8. Bibcode1972Natur.237...82J. doi:10.1038/237082a0. PMID 4555447.

[4] Fiers W, Contreras R, Duerinck F, et al (April 1976). "Complete nucleotide sequence of bacteriophage MS2 RNA: primary and secondary structure of the replicase gene". Nature260 (5551): 500–7. Bibcode 1976 Natur.260..500F. doi:10.1038/260500a0.PMID 1264203.

[5] Maxam AM, Gilbert W (February 1977). "A new method for sequencing DNA". Proc. Natl. Acad. Sci. U.S.A. 74 (2): 560–4. Bibcode 1977PNAS...74..560M.doi:10.1073/pnas.74.2.560. PMC 392330. PMID 265521.

[6] Gilbert, W. DNA sequencing and gene structure. Nobel lecture, 8 December 1980.

[7] Gilbert W, Maxam A (December 1973). "The Nucleotide Sequence of the lac Operator".Proc. Natl. Acad. Sci. U.S.A. 70 (12): 3581–4. Bibcode 1973PNAS...70.3581G.doi:10.1073/pnas.70.12.3581. PMC 427284. PMID 4587255.

[8] Sanger F, Coulson AR (May 1975). "A rapid method for determining sequences in DNA by primed synthesis with DNA polymerase". J. Mol. Biol. 94 (3): 441–8.doi:10.1016/0022-2836(75)90213-2. PMID 1100841.

[9] Sanger F, Nicklen S, Coulson AR (December 1977). "DNA sequencing with chain-terminating inhibitors". Proc. Natl. Acad. Sci. U.S.A. 74 (12): 5463–7. Bibcode1977PNAS...74.5463S. doi:10.1073/pnas.74.12.5463. PMC 431765.PMID 271968.

[10] Sanger F. Determination of nucleotide sequences in DNA. Nobel lecture, 8 December 1980.

[11] 10.Graziano Pesole; Cecilia Saccone (2003). Handbook of comparative genomics: principles and methodology. New York: Wiley-Liss. pp. 133. ISBN 0-471-39128-X.

[12] Smith LM, Fung S, Hunkapiller MW, Hunkapiller TJ, Hood LE (April 1985). "The synthesis of oligonucleotides containing an aliphatic amino group at the 5' terminus: synthesis of fluorescent DNA primers for use in DNA sequence analysis". Nucleic Acids Res. 13 (7): 2399–412. doi:10.1093/nar/13.7.2399. PMC 341163. PMID 4000959.

[13] Base-calling for next-generation sequencing platforms — Brief Bioinform". Retrieved 2011-02-24.

[14] Murphy, K.; Berg, K.; Eshleman, J. (2005). "Sequencing of genomic DNA by combined amplification and cycle sequencing reaction". Clinical chemistry 51 (1): 35–39.doi:10.1373/clinchem.2004.039164. PMID 15514094. edit

[15] Sengupta, D. .; Cookson, B. . (2010). "SeqSharp: A general approach for improving cycle-sequencing that facilitates a robust one-step combined amplification and sequencing method". The Journal of molecular diagnostics : JMD 12 (3): 272–277.doi:10.2353/jmoldx.2010.090134. PMC 2860461. PMID 20203000. edit

[16] Richard Williams, Sergio G Peisajovich, Oliver J Miller, Shlomo Magdassi, Dan S Tawfik, Andrew D Griffiths (2006). "Amplification of complex gene libraries by emulsion PCR".Nature methods 3 (7): 545–550. doi:10.1038/nmeth896. PMID 16791213.

[17] Hall N (May 2007). "Advanced sequencing technologies and their wider impact in microbiology". J. Exp. Biol. 210 (Pt 9): 1518–25. doi:10.1242/jeb.001370.PMID 17449817.

[18] Church GM (January 2006). "Genomes for all". Sci. Am. 294 (1): 46–54.doi:10.1038/scientificamerican0106-46. PMID 16468433.

[19] Schuster, Stephan C. (2008). "Next-generation sequencing transforms today's biology".Nature methods (Nature Methods) 5 (1): 16–18. doi:10.1038/nmeth1156.PMID 18165802.

[20] Brenner, Sidney; Johnson, M; Bridgham, J; Golda, G; Lloyd, DH; Johnson, D; Luo, S; McCurdy, S et al. (2000). "Gene expression analysis by massively parallel signature sequencing (MPSS) on microbead arrays". Nature Biotechnology (Nature Biotechnology)18 (6): 630–634. doi:10.1038/76469. PMID 10835600.

[21] Schuster SC (January 2008). "Next-generation sequencing transforms today's biology". Nat. Methods 5 (1): 16–8. doi:10.1038/nmeth1156. PMID 18165802.

[22] Mardis ER (2008). "Next-generation DNA sequencing methods". Annu Rev Genomics Hum Genet 9: 387–402. doi:10.1146/annurev.genom.9.081307.164359.PMID 18576944.

[23] Valouev A, Ichikawa J, Tonthat T, et al. (July 2008). "A high-resolution, nucleosome position map of C. elegans reveals a lack of universal sequence-dictated positioning".Genome Res. 18 (7): 1051–63. doi:10.1101/gr.076463.108. PMC 2493394. PMID 18477713.

[24] Human Genome Sequencing Using Unchained Base Reads in Self-Assembling DNA Nanoarrays. Drmanac, R. et. al. Science, 2010, 327 (5961): 78-81,

[25] Genome Sequencing on Nanoballs Porreca, JG. Nature Biotechnology, 2010, 28:(43-44)

[26] Human Genome Sequencing Using Unchained Base Reaads in Self-Assembling DNA Nanoarrays, Supplementary Material. Drmanac, R. et. al. Science, 2010, 327 (5961):78-81, Complete Genomics Press release, 2010

[27] Hanna GJ, Johnson VA, Kuritzkes DR, et al (1 July 2000). "Comparison of Sequencing by Hybridization and Cycle Sequencing for Genotyping of Human Immunodeficiency Virus Type 1 Reverse Transcriptase". J. Clin. Microbiol. 38 (7): 2715–21. PMC 87006.PMID 10878069.

Section 2

Applications of DNA Sequencing

3

DNA Sequencing and Crop Protection

Rosemarie Tedeschi
University of Torino – DIVAPRA –
Entomologia e Zoologia applicate all'Ambiente "C. Vidano"
Italy

1. Introduction

Many plant pathogenic organisms (e.g. viruses, fungi, bacteria and phytoplasmas) as well as plant pests (e.g. insects and mites) cause serious and widespread diseases and injuries to different cultivated plants with heavy economic consequences for the growers.

Crop management to reduce damage by diseases and pests is based on integrated control strategies involving exclusion, eradication, and protection. In the recent years the choice among the different options is driven not only by the overall costs versus the increased yield and/or quality, but also by the energy consumption, time taken, environmental impact and overall sustainability.

One of the crucial step in crop protection is the rapid, accurate and reliable plant pathogen/pest detection, identification and quantification. It allows to control the spread of the diseases/pests by screening vegetal propagative material and to implement quarantine regulations. Moreover pathogen/pest detection and identification are fundamental for epidemiological studies and for the design of new control strategies.

Traditionally, the most used approach to identify plant pathogens relied with visual inspection of symptoms usually followed by laboratory analyses based on morphological identification using microscopy and isolation and culturing of the organisms. In some cases these methods are still used, but actually these conventional approaches require skilled and specialized expertise which often takes many years to acquire, are laborious, time consuming and not always sensitive and specific enough. Moreover closely related organisms may be difficult to discriminate on morphological characters alone, symptoms are not always specific and not all the microorganisms are culturable in vitro. For all these reasons in the last decades much effort has been devoted to the development of novel methods, in particular nucleic acid based molecular approaches, for detecting and identifying plant pathogens and pests (Lopez et al., 2003; McCartney et al., 2003; Alvarez, 2004; Lievens & Thomma, 2005). The use of DNA-based methods derives from the premise that each species of pathogen carries unique DNA or RNA signature that differentiates it from other organisms. Knowing the pathogen/pest nucleic acid sequence enables scientists to construct oligos to detect them. These oligos are at the basis of many highly specific analytical tests now on the market (Louws et al., 1999).

The present chapter will describe the routine applications of DNA-based technology in different aspects of crop protection. It will highlight the perspectives of innovative approaches, based on nucleotide sequences, in providing new sustainable and environmentally-friend strategies for disease and pest management. Moreover the role of next-generation sequencing (NGS) technology in supplying these innovative approaches will be outlined.

2. Molecular tools for detection and identification of plant pathogens and pests

Many molecular techniques based on the hybridization or amplification of target sequences, have been developed for many plant pathogenic organisms and pests as well. A great and indisputable improvement in molecular diagnostic was achieved with the introduction of the polymerase chain reaction (PCR) in the mid 1980s (Henson & French, 1993). PCR technology is now widely used for plant pathogen detection because it is rapid, specific and highly sensitive. Even RNA-based viruses can be detected by PCR, by first reverse-transcribing the RNA into DNA. The specificity, that is the ability to distinguish closely related organisms, depends on the designing of proper primers and improvements in sequencing technologies are making the selection of reliable PCR primers routine (Schaad & Frederick, 2002; Schaad et al., 2003). Highly conserved gene regions are often the target for designing primers. Often, to select the target sequences to be used in nucleic acid diagnostic, the same known gene is isolated and sequenced from target and non-target pathogens. Then regions of the sequence that are different are used for primer design. Alternatively, randomly selected DNA fragments are used. With the advancements in high-throughput DNA sequencing, more and more genomes of plant pathogens are available (http://cpgr.plantbiology.msu.edu) and a huge number of nucleotide sequences has been deposited in databases, i.e. GeneBank. Searching in these databases the sequences of a particular organism and screening them for their feasibility as potential targets in a PCR-based assay of a particular organism makes the primer design more and more rapid and precise.

In viruses the most common target is the coat protein gene. In bacteria, phytoplasmas and fungi very often the DNA encoding the ribosomal RNA (rDNA) is used, that occurs as a repeated, structured unit consisting of relatively conserved ribosomal RNA (rRNA) subunit genes (16S, 23S and 5S in the case of prokaryotes and 8S, 5.8S and 28S in the case of fungi) which are separated by internal transcribed spacers (ITS) with relatively high variability. This allows to design primers binding to conserved regions and amplify variable regions on sequences that are conserved between species. These variable regions can be used for species identification (White et al., 1990). Usually the most used techniques are conventional specific PCR, nested-PCR to reach high sensitivity or PCR followed by restriction fragment length polymorphism (PCR+RFLP) to reach high specificity. However, ribosomal sequences do not always reflect sufficient sequence variation to discriminate between particular species (Tooley et al., 1996). Therefore, but also to corroborate discrimination based on ITS sequences, other housekeeping genes are becoming more intensively studied, including beta-tubulin (Fraaije et al., 1999; Hirsch et al., 2000), actin (Weiland & Sundsbak, 2000), elongation factor 1-alpha (O'Donnell et al., 1998; Jiménez-Gasco et al., 2002), and mating type genes (Wallace & Covert, 2000; Foster et al., 2002).

Besides detection and identification, quantification of a pathogen is fundamental in crop protection because it is related to the risk of disease development and economic loss and thus directly linked to disease management decision. For this reason real-time PCR, allowing amplified DNA quantification using fluorescent dyes, and guaranteeing reduced risk of sample contamination avoiding post-PCR sample processing, is one of the most rapid species-specific detection techniques currently available.

We should also consider that plants can be infected by different pathogens at the same time, thus technologies able to detect multiple pathogens simultaneously are required. Multiplex PCR, using several PCR primers in the same reaction, allows these kind of analyses saving time and reducing costs. However multiplex PCR can be difficult to develop because the products from different targets need to be different sized to ensure that they can be distinguished from one another and they must all be amplified efficiently using the same PCR conditions (Henegariu et al., 1997). Also real-time PCR allows the simultaneous detection of different targets by using probes with different fluorescent reporter dye (Weller et al. 2000, Boonham et al., 2000). Microarray technology is an emerging tool in crop protection for unlimited multiplexing analyses (Fessehaie et al., 2003; Lèvesque et al., 1998; Lievens et al., 2003; Uehara et al., 1999; Mumford et al. 2006; Boonham et al. 2007), but nowadays existing microarray methods are still complex and relatively insensitive, and a widely accepted diagnostic format has yet to be adopted.

PCR-based techniques are also being used in an insect pest management context to identify insect pests and insect biotypes, to understand population structure, tritrophic interactions and insect-plant relationships (Caterino et al., 2000; Heckel, 2003; Gariepy et al., 2007). For this purpose the ribosomal and mitochondrial regions have been proven to be highly informative. In particular the cytochrome c oxidase I and II (COI and COII) and the 16S and 12S subunits of mitochondrial DNA as well as the internal transcribed spacer regions (ITS1 and ITS2), the 18S and the 28S subunits of the rDNA are used (Gariepy et al., 2007).

However DNA-based identification of species should not be considered as a replacement for the traditional methods, but it is a great help in all the situations in which morphological characters are not helpful such as in the case of young stages or cryptic species. In particular it can help in the study of known or putative vectors of plant pathogens giving innovative contributions to faunal studies and insect vector monitoring and allowing more rational control strategies (Bertin et al., 2010a; Bertin et al., 2010b; Cavalieri et al., 2008; Saccaggi et al., 2008; Tedeschi & Nardi, 2010).

Molecular techniques have also been used to facilitate ecological studies on parasitoids and predators in biological control programs (reviewed by Gariepy et al., 2007; Greenstone, 2006), while microsatellite analysis can be used to separate strains and evaluate genetic diversity of natural enemy populations. All these information may be able to predict natural enemy host range, climatic adaptability and other important biological traits that can be used in the selection of efficient candidate biological control agents (Unruh & Woolley, 1999).

Anyhow all these molecular detection methods such as PCR and real-time PCR, rely on previous knowledge of the pathogen/pest for design of sequence-specific primers.

The advent of the second generation sequencing affords unique opportunity to directly detect, identify and discover pathogens without requiring prior knowledge of the pathogen macromolecular sequence. This technology is unbiased and allows to consider, on the whole, bacteria, viruses, fungi and parasites (Quan et al., 2008).

This approach, already applied in medical diagnostic, starts to be regarded in phytopathology. As correctly stated by Studholme et al. (2011), efficient sample preparation methods, bioinformatics pipelines to efficiently discriminate pathogens and host sequences and functionally associating candidate sequences with disease causation should be developed before using high-throughput sequences as a routine diagnostic tool. Moreover plant samples always contain a complex microbiota that alone cannot reveal the causal agents of a disease and biological knowledge are fundamental for a good interpretation of the results.

The total nucleic acid from the sample of interest has the advantage that it avoids introducing bias and is suitable for a variety of substrate, but the elimination of host nucleic acid is very critical to increase the pathogen signal towards the threshold for detection.

The huge number of sequences obtained should be trimmed and filtered to remove low-quality sequences and reads containing any primer sequences. Then the sequences are compared against databases of known sequences using a pair-wise sequence similarity method. If no similar sequences are found, other methods can be used, such as the ones that separate sequences into clusters based on their physical characteristics (Abe et al., 2003; Chan et al., 2008).

If a new pathogen is identified on the basis of sequences alone, problems could be occur for its naming and taxonomy, because Koch's postulate are rarely completed, thus maybe a prefix or suffix can highlight this status (Studholme et al., 2011).

3. Application of DNA barcoding in crop protection

DNA barcoding is a universal typing system to ensure rapid and accurate identification of a broad range of biological specimens. This technique allows the species characterization of organisms using a short DNA sequence from a standard and agreed-upon position in the genome. This concept was proposed by Hebert et al. in 2003 with the description of the first marker as a "barcode", the mithocondrial COI gene, for species identification in the animal kingdom. The DNA target sequence should be identical among the individuals of the same species, but different between species, extremely robust with highly conserved priming sites and highly reliable DNA amplifications and sequencing, phylogenetically informative and short enough to have lower processing costs and to allow amplification of degraded DNA (Valentini et al., 2008). However the perfect DNA target region does not exist and more than one marker have been proposed. The COI region has been almost widely accepted for barcoding animals because of its generally conserved priming sites and third-position nucleotides with a greater incidence of base substitutions than other mitochondrial genes. Moreover the evolution of this gene is rapid enough to allow the discrimination of not only closely allied species, but also phylogeographic groups within a single species (Cox & Herbert, 2001; Wares & Cunningham, 2001). On the contrary in plants the mtDNA has low substitution rates and a rapidly changing gene content and structure, which makes COI unsuitable for barcoding in plants (Wolfe et al., 1987). For this reason two regions of chloroplast DNA, ribulose-bisphosphate carboxylase (rbcL) and maturase K (matK) have been recommended for initiating the barcoding process of plant species (CBOL Plant

Working group, 2009; Consortium for the barcode of life, 2009). Due to the only 70% species discriminatory power, additional loci need to be used in this field (Vijayan & Tsou, 2010). On the contrary in the case of fungi, the ITS of nuclear DNA (nrDNA) has recently been proposed as the official primary barcoding marker (Seifert, 2009), while no standard regions have been found yet for viruses, bacteria and phytoplasma, but studies are ongoing (Contaldo et al., 2011).

The primary intent of DNA barcoding was to use large-scale screening of one or a few reference genes in order to assign unknown individuals to species, and to enhance discovery of new species (Hebert et al. 2003; Stoeckle, 2003), so this approach was extensively used by taxonomists and ecologists in particular for biodiversity studies in spite of limitations and pitfalls (Valentini et al., 2009; Moritz & Cicero, 2004). Today new applications are emerging and the use of DNA barcoding has been proposed for a new approach in plant protection, that is for the rapid and precise identification of invasive alien species (IAS). Non-indigenous pests, and the plant pathogens they harbor are a serious threat for agriculture and forestry with huge ecological and economical consequences. IAS occur in all major taxonomic groups, including viruses, fungi, bacteria, nematodes and insects. All these organisms are able to reproduce and spread, causing massive environmental damage also because in the new environment often no natural specific predators or parasitoids are present. Early detection and rapid response are most cost-effective and more likely to succeed than action after a species has become established. For this reason inspections and quarantines are key factors of prevention. As previously stated the development of accurate identification tools for plant pathogens and pests is fundamental for plant health in agriculture but at the same time the taxonomic knowledge available to identify harmful quarantine organisms via their visual characteristics is gradually decreasing. DNA barcoding, using standardized protocols, will easily allow to identify specimens comparing their sequences against a database of known sequences from identified specimens. Based on DNA, this system is applicable to all life stages even those that cannot be identified by conventional means such as eggs and larvae of insects. This approach does not require extensive knowledge of traditional morphological taxonomy and a minimum of technical expertise is sufficient (Floyd et al., 2010).

Several major research initiatives have already started to assemble reference libraries of DNA barcodes of pest species in quarantine. The International Plant Protection Convention (IPPC) agreed that barcoding currently has a clear potential as a diagnostic protocol.

In New Zealand, since 2005, DNA barcoding using COI sequence has been employed routinely for the highest risk insect species, fruit flies and lymantriid moths. In particular, case studies with the pink gypsy moth *Lymantria mathura* Moore, the yellow peach moth *Canocethes punctiferalis* (Guenée) and the fall web worm *Hyphantria cunea* (Drury) demonstrated the effectiveness of DNA barcoding for border diagnostics. Moreover barcoding improved the chances of finding new organisms, even those that may not have been identified already as high risk to New Zealand (Armstrong, 2010).

In Europe, a project financed by the 7th Framework Program of the European Union and called QBOL (http://www.qbol.org) aims for the development of a new diagnostic tool using DNA barcoding to identify quarantine organisms in support of plant health. QBOL is a project that makes collections harboring plant pathogenic quarantine organisms (arthropods, bacteria, fungi, nematodes, phytoplasma and viruses) available (Bonants et al.,

2010). Informative genes from selected species on the EU Directive and EPPO lists are DNA barcoded from vouchered specimens and the sequences, together with taxonomic, phylogenetical and phytosanitary features, will be included in an internet-based Q-bank database (http://www.q-bank.eu) (Bertaccini et al., 2011). Therefore DNA barcoding will supply not only a strengthen link between traditional and molecular taxonomy as a sustainable diagnostic resource, but also a diagnostic tool for plant health.

Moreover, considering its application in identification and interspecific discrimination of phylogeographic groups within a single species, important advices can be achieved by IAS. Indeed, it will be possible to assess from where exotic specimens originated and get data on the natural host range, climatic range, and potential biocontrol agents, all crucial information for the development of successful control programs (Valentini et al., 2009).

Another practical application of DNA barcoding could support crop protection strategies: DNA barcoding is also a useful tool for searching for candidates of biological control agents and evaluating their potential risks. Normally, searching and screening of control agents require long-term feeding experiments, while DNA barcoding allows to identify and select control agents based on their gut contents (Jinbo et al., 2011; Neumann et al., 2010), detecting prey DNA from the gut content or feces of the predators (Symondson, 2002; Valentini et al., 2008).

Anyhow whatever purpose the DNA barcoding is used for, very good sequence reference datasets for groups with well worked out taxonomy are important key factors.

Due to the evolution of DNA sequencing technology and to the possibility to have nowadays an easy access to sequencing via companies, the DNA-barcoding had a general great impulse and spreading. When using the classical sequencing approach via capillary electrophoresis, environmental samples require an additional step of cloning the different amplified DNA fragments into bacteria, followed by sequencing hundreds or thousands of clones to reveal the full complexity of those samples. Such a cloning step is both expensive and time consuming, thus limiting large-scale applications. New DNA sequencing technologies bypassing the cloning step have recently been developed, opening the way to applying large-scale DNA barcoding studies to environmental samples.

4. Molecular crop protection using recombinant DNA techniques

Conventional breeding techniques which are based on processes of crossing, back crossing and selection have been used since the beginning of agriculture. With these methods, that have been used for hundreds of year, farmers could obtain plants with faster growth, higher yields, pest and disease resistance, larger seeds, or sweeter fruits. In spite of the undoubted good results reached, conventional breeding proved to be time consuming and could not stand the pace with the rapid co-evolution of pathogenic micro-organisms and pests.

A great improvement was achieved in the last 30 years by tissue cultures and micropropagation for the production of disease-free, high quality planting material and for rapid production of many uniform plants. Moreover the rapid advances in molecular biology technologies led to new knowledge for cloning and sequence analysis of the genomic components of plants and the relative pathogens and pests. Comparative mapping across species with genetic markers, and objective assisted breeding after identifying candidate genes or chromosome regions for further manipulations brought to the development and application of recombinant DNA techniques in crop pest management.

The introduction in plants (transgenic plants) of new genes from other organisms (native or modified forms) allows to develop resistant plants against insects or pathogens (mainly viruses) or more tolerant plants towards herbicides. In contrast to conventional breeding, which involves the random mixing of tens of thousands of genes present in both the resistant and susceptible plant, recombinant DNA technology allows the transfer of only the resistance gene to the susceptible plant and the preservation of valuable economic traits.

The transfer of new genes in plants have been obtained mainly using *Agrobacterium*-mediated DNA transfer or by particle bombardment, electroporation and polyethylenglycol permeabilisation (Fisk & Dandekar, 1993). Strategies using recombinant DNA allowed to obtain different degrees of resistance against viruses, fungi, bacteria and insects.

The engineering of virus resistance into plants is considered one of the most important way to control these pathogens, due to the fact that no direct chemical control is available and the control of the insect vectors or the use of certified virus free material are not overly successful. The first approach, based on the pathogen-derived resistance (PDR) (Sanford & Johnson, 1985) was applied to many plant species. It entails the introduction of a virus-derived gene sequence into the genome of the host plant, then the products of these sequences (mRNA and proteins) interfere with the specific stages in the viral infection cycle such as viral replication or spread. The use of the viral coat protein (CP) as a transgene gave the most important results, with delay and attenuation of symptoms or complete immunity. Powell et al. (1986) first introduced, in this way, resistance against Tobacco mosaic virus (TMV) in tobacco plants then many other researchers obtained several engineered crops for commercial cultivation such as tomato resistant to TMV, Tomato mosaic virus (ToMV) (Sanders et al., 1992), Cucumber mosaic virus (CMV) (Fuchs et al., 1996), potato resistant to Potato virus X (PVX) (Jongedijk et al. 1992), Potato virus Y (PVY) (Malnoe et al., 1994) and Potato leafroll virus (PLRV) (Presting et al., 1995). Moreover tobacco containing the CP genes for three viruses, Tomato spotted wilt virus (TSWV), Tomato chlorotic spot virus (TCSV), and Groundnut ringspot virus (GRSV) developed resistance to all of them (Prins et al., 1995). Alternatives to CP-mediate protection are the introduction of sequences coding for non-structural viral proteins, the so called replicase-mediated protection, the movement-protein-mediated resistance and the REP protein-mediated resistance to single stranded DNA that proved to be effective for controlling virus diseases in a wide range of crops (Prins et al., 2008). Moreover new and important opportunities came from the RNA-mediated resistance, founded on the post-transcriptional gene silencing process called also RNA interference (RNAi) or RNA silencing. It is based on a plant natural antiviral mechanism that acts against a virus by degrading its genetic material in a nucleotide sequence specific manner that will be deepened in the following paragraph.

Strategies proposed to control fungal and bacteria diseases by transgenic plants are similar, because of their similarity in modes of pathogenesis and plant response. As a defense strategy against the invading pathogens (fungi and bacteria) the plants accumulate low molecular weight proteins which are collectively known as pathogenesis-related (PR) proteins. Several transgenic crop plants with increased resistance to fungal pathogens are being raised with genes coding for the different compounds such as the glucanase or chitinase enzymes that degrade the cell wall of many fungi (Grover & Gowthaman, 2003), or lysozyme and cercopsins that cause the hydrolytic cleavage of cell wall murein (Strittmatter & Wegener, 1993) and the formation of ion channels in the bacterial membrane with the leakage of cell components respectively (Florack et al., 1995). For both fungi and bacteria also the use of phytoalexins gave good results together with many other strategies well resumed by Grover & Gowthaman (2003).

Insect resistant crops are one of the major success of recombinant DNA technology in agriculture. Cotton resistant to lepidopteran larvae and maize resistant to both lepidopteran and coleopteran larvae are nowadays widely used and allowed to reduce the use of pesticides and the overall production costs (Gatehouse, 2008).

The soil bacterium *Bacillus thuringiensis* (Bt) is one of the most important source of genes encoding insecticidal proteins, which are accumulated in the crystalline inclusion bodies produced by the bacterium on sporulation (Cry proteins, Cyt proteins) or expressed during bacterial growth (Vip proteins). These are specific to particular groups of insect pests (mainly Lepidoptera and Coleoptera), and are not harmful to other useful insects. Moreover different Bt strains show different specificity against different insect pests. For this reasons induced resistance using Bt genes is one of the first application of crop protection with recombinant DNA technology. Expression of Bt genes in tobacco and tomato provided the first example of genetically engineered plants for insect resistance (Barton et al., 1987; Vaeck et al., 1987). Then after, hundreds of field trials with different transgenic crops with Cry genes (e.g. cotton, rice, maize, potato, tomato, cauliflower, cabbage, etc.) have proven the efficacy of this approach in controlling important insect pests with a consequent impressive reduction in pesticide use (Gatehouse, 2008).

The possibility of insects to evolve resistance to transgenic Bt crops has been expected by many researchers due to the widespread of Bt transgenic crops and to the prolonged exposure to Bt toxins (Tabashnik et al., 2008), but despite a few documented cases (Bates et al., 2005), most insect pest populations are still susceptible. To preserve this susceptibility, some agronomical devices are crucial, such as the refuge strategy, that is the presence near Bt crops of host plants without Bt to promote survival of susceptible pests (Tabashnik et al., 2008). At the same time, to reduce the risk of insect resistance, new strategies are proposed such as the expression of multiple Cry genes in transgenic crops (Zhao et al., 2003) or the use of completely different genes.

Certain genes from higher plants were also found to result in the synthesis of products possessing insecticidal activity. For instance many plants produce protease inhibitors that are small proteins which inhibit the proteases present in insect guts or secreted by microorganism, causing a reduction in the availability of amino acids necessary for their growth and development (De Leo et al., 2002). One of the examples is the Cowpea trypsin inhibitor gene (CpTi) which was introduced into tobacco, potato, and oilseed rape for developing transgenic plants (Boulter et al., 1989). It was observed that the insecticidal protein was a trypsin inhibitor that was capable of destroying insects belonging to the orders Lepidoptera, Orthoptera, etc. while it has no effect on mammalian trypsin, hence it is non-toxic to mammals.

Another approach that maybe will have more chances in the future is the production of transgenic plants expressing genes encoding lectins (Rao et al., 1998) considering that many plant lectins have strong insecticidal properties causing a deterrent activity towards feeding and oviposition behavior (Murdock & Shade, 2002; Michiels et al., 2010).

The increasing availability of DNA sequence information allow to discover genes and molecular markers associated with different agronomic traits creating new opportunities for crop improvement.

The complete genome sequencing can be considered the highest level of genetic markers, providing the opportunity to explore genetic diversity in crops and their wild relatives on a much larger scale than with the earlier technologies. Nowadays the complete genome sequences are becoming available for many plants opening new important opportunities for the creation of transgenic plants resistant or tolerant to pathogens and pests.

5. Crop protection using RNA interference

RNA interference (RNAi) is a naturally occurring regulatory mechanism which causes sequence specific gene silencing. Actually it refers to diverse RNA-based processes that all result in sequence-specific inhibition of gene expression either at the transcriptional, post-transcriptional or translation levels.

This phenomenon was first observed in plants as a unexpected result of attempt to make the color of petunia blooms more purple. Trying to overexpress the production of anthocyanin pigments introducing exogenous transgenes, flowers showed, instead of deepen flower colour, variegated pigmentation, with some lacking pigment altogether (Napoli et al., 1990). This mechanism, with apparent silencing of both endogenous and exogenous genes, was indicated as post-transcriptional gene silencing (PTGS) and termed "co-supression". Some years later RNAi was described in the nematode *Caenorhabditis elegans* (Maupas) as a response to double-stranded RNA (dsRNA) injected into the body cavity, which resulted in sequence-specific gene silencing expressed in distal tissues and in the progeny (Fire et al., 1998). RNAi could be induced also by feeding the worms on the engineered *Escherichia coli* producing dsRNA against the target gene (Fire et al., 1998; Timmons & Fire, 1998; Timmons et al., 2001), by in vitro transcription of transgene promoters (Tavernarakis et al., 2000), or simply by soaking the worms in a solution containing the dsRNA (Tabara et al., 1998). Then Ketting & Plasterk (2000) demonstrated that the co-suppression recorded in plants and the phenomena observed in nematodes are based on the same mechanism and the term RNAi was adopted.

It is now commonly accepted that it has evolved as a mechanism for cells to eliminate foreign genes; indeed it is considered to be an evolutionary ancient mechanism for protecting organisms from viruses (Waterhouse et al., 2001). Many viruses have RNA as their genetic material and it is widely assumed that ds-RNA is generated by viral RNA polymerases either as an intermediate in genome replication (RNA viruses) or as an erroneous product due to converging bidirectional transcription (DNA viruses) (Kumar & Carmichael, 1998; Ketting et al., 2000, Jacobs & Langland, 1996). All the pluricellular organism hosts are able to recognize dsRNA and then activate the RNAi defensive mechanism. An endoribonuclease called Dicer cleaves this dsRNA into short dsRNA fragments called small interfering RNA (siRNA) about 20-25 nucleotides long. Then a multiprotein complex, the RNA-Induced Silencing Complex (RISC), uses the siRNA as a template for recognizing and destroying single stranded RNA with the same sequence, such as mRNA copies used by the virus to direct synthesis of viral protein. It was also demonstrated that some animal and plant viruses are able to produce proteins to suppress host-mediated RNA silencing allowing viral spread within the host (Li et al., 2002; Ding et al., 2004).

This kind of homology-dependent gene silencing appears to be evolutionarily conserved in all eukaryotic taxa and it was recorded in plants, fungi, invertebrate, including insects, trypanosomes, planaria and hydra, and vertebrates (Bosher & Labouesse, 2000). Even if some specifics of the silencing mechanism may differ between kingdoms, dsRNA seems to be the universal initiator of RNAi.

Since its discovery, RNAi has become a remarkably potent technology to knockdown gene functions in a wide range of organisms representing a valuable tool for functional genomics; moreover it has also found a great potential application in agriculture and in particular it offers a new approach to crop protection. Due to its high specificity, it can be considered a species-specific tool against pests. This phenomenon has been studied in various host-pest systems and efficiently used to silence the action of the pest. New RNAi based strategies have been proposed in last years for conferring plant resistance to phytopathogenic viruses, bacteria, fungi and nematodes and some have been used in commercial agriculture (Escobar et al., 2001; Huang et al., 2006; Gonsalves, 1998). Moreover particular attention was devoted to RNAi efforts targeting insects.

Considering its ancestral role of conferring protection against viral infection, the possibility to manipulate genetically the plants in order to obtain transgenic resistance was supported. Different plant-virus systems have been investigated for inducing immunity by means of RNAi technology and they are well resumed by Wani et al. (2010) and Wani & Sanghera (2010). Very recently Zhang et al. (2011) were able to achieved robust resistance to multiple viruses (Alfalfa mosaic virus, Bean pod mottle virus, and Soybean mosaic virus) with a single dsRNA-expressing transgene in soybean plants opening new interesting perspectives in practical viruses control. It has also been assessed that plant viruses have evolved mechanism to overcome the host antiviral silencing by encoding diverse viral suppressors of RNA silencing (VSRs). These suppressors enhance virus accumulation in the inoculated protoplasts, promote cell-to-cell virus movement in the inoculated leaves, facilitate the phloem-dependent long-distance virus spread, often cause more severe disease (Díaz-Pendón & Ding, 2008; Li & Ding, 2006).

In the case of bacteria very interesting efforts were carried out to generate resistance to crown gall disease caused by the soil bacterium *Agrobacterium tumefaciens* (Smith & Townsend) that causes serious economic losses in perennial crops worldwide. Escobar et al. (2001) obtained transgenic *Arabidopsis thaliana* L. (Heyn) and *Lycopersicon esculentum* L. plants expressing two self-complementary RNA constructions designed to initiate RNAi of tryptophan monooxygenase (iaaM) and isopentenyl transferase (ipt) genes. Expression of these two oncogenes causes the overproduction of the plant hormones auxin and cytokinin resulting in the formation of wild type tumors. The resultant transgenic lines retained susceptibility to *Agrobacterium* transformation but were in some cases highly refractory to tumorigenesis, providing functional resistance to crown gall disease. This procedure could be exploited to develop broad spectrum resistance in ornamental and horticultural plants which are susceptible to crown gall tumorigenesis.

RNAi mechanism has been proposed also to improve plant resistance to fungi. The first gene silencing phenomena in a fungus was described in 1992 by Romano and Macino in *Neurospora crassa* Shear & Dodge. Later on many other fungal species were studied and suppression of gene expression by dsRNA-expressing plasmid or related-systems has been shown in various Ascomycota such as *Venturia inaequalis* (Cooke) Winter (Fitzgerald et al., 2004), *Aspergillus nidulans* (Eidam) Winter (Hammond & Keller, 2005), *Magnaporthe oryzae* Couch (Kadotani et al., 2003), *Aspegillus fumigatus* Fresenius (Mouyna et al., 2004), Basidiomycota such as *Cryptococcus neoformans* (Sanfelice) Vuillemin (Liu et al., 2002) and Zygomicota.

After the first demonstration of RNAi mechanism in the free-living nematode *C. elegans*, the possibility that dsRNA mediates gene silencing also in other worm species was mentioned by Fire et al. (1998). Important advances have been made in the biotechnological application of RNAi towards plant parasitic nematode control. RNAi silencing of a gene that plays a key role in the development of the nematode, either directly or indirectly, can adversely affect the progression of pathogenesis. Genes that are good targets for this technology are likely to be nematode specific and have sequence conservation with orthologues from related species to maximize the spectrum of resistance (Karakaş, 2010). However RNAi in plant parasitic nematodes by injection or soaking is not feasible, because they are too small to be microinjected with dsRNA and do not normally ingest fluid until they have infected a host plant (Bakhetia et al., 2005). But in 2002 Urwin et al. induced the nonfeeding preparasitic second stage (J2) of two cyst nematodes *Globodera pallida* (Stone) and *Heterodera glycines* Ichinohe (which parasitize potato and soybean respectively) to take up dsRNA from a solution containing the neuroactive compound octopamine. Subsequently they observed a shift from the normal female/male ratio of 3:1 to 1:1 by 14 days postinfection (dpi) targeting cysteine proteinases and a reduction of the number of parasites targeting C-type lectine gene. Later, other two stimulation reagents, resorcinol and serotonin, were also used to induce dsRNA uptake by J2s much more effectively (Rosso et al., 2005). In the last years RNAi assays on sedentary forms followed one another and included root knot nematodes (e.g. Meloidogyne), cyst nematodes (e.g. Heterodera and Globodera), and the migratory pine nematodes *Bursaphelenchus xylophilus* (Steiner et Buhrer) Nickle. All these studies, that are well summarized by Lilley et al. (2007) and Rosso et al. (2009), referred to the successful targeting of a number of genes, expressed in a range of different tissues and cell types, for silencing in various plant parasitic nematodes. Plants that express dsRNA or hairpin RNA in the nematode feeding cells provide a new option for creation of resistant varieties. Plant parasitic nematodes can be found deep enough in the soil to escape the action of pesticides. The development of trapping plants that attract nematodes and block their development through the expression of dsRNA that target essential nematode genes, may be an efficient way of reducing nematode populations in the field (Rosso et al., 2009).

In the field of entomology, the RNAi technique has already widely applied for different kind of researches, to improve the knowledge of RNAi mechanisms *in primis*, but also to study gene function as well as expression and regulation of gene cascades (Price & Gatehouse, 2008; Whyard et al., 2009; Bellés, 2010; Mito et al., 2011). Successful results were obtained in preventing bee mortality by Israeli Acute Paralysis Virus Disease (IAPV) infection under natural beekeeping conditions (Hunter et al., 2010), but many efforts focused on the possibility to control insect pests by means of RNAi. For these studies different insect models were used, such as *Drosophila melanongaster* Meigen (Misquitta & Paterson, 1999; Dzitoyeva et al., 2001; Schmid et al., 2002), *Tribolium castaneum* (Herbst) (Bucher et al., 2002; Tomoyasu & Denell, 2004; Tomayasu et al., 2008), honeybees (Amdam et al., 2003), *Acyrthosiphon pisum* (Harris) (Mutti et al., 2006; Jaubert-Possamai et al., 2007), the green peach aphid *Myzus persicae* Sulzer (Pitino et al., 2011), the western corn rootworm *Diabrotica virginifera virginifera* (LeCompte) (Baum et al., 2007; Alves et al., 2010). In almost all the experiments carried out, dsRNA was injected directly in the organism, but this kind of dsRNA delivery is not feasible in the protection of crops against phytophagous insects.

The big obstacles to make RNAi approach extensively feasible for crop protection is related to the selection of target sequences, to the production of big quantities of dsRNA to be released, and to the delivery systems of dsRNAs.

Concerning the selection of target sequences, up to now RNAi application tooks advantages from already known genes while cDNA libraries allowed to identify novel genes (Turner et al., 2006; Baum et al., 2007).

The recent availability for many organisms of the full genome sequences makes possible, at least theoretically, the silencing of every single gene to evaluate how the organism reacts. A great improvement in the selection of target genes can be given by the next-generation sequencing (NGS) technologies, with the direct sequence of the cDNA generated from messenger RNA (RNA-seq) (Wang et al., 2009; Haas & Zody, 2010). This new method, called also "Whole Transcriptome Shotgun Sequencing", uses recently developed high-throughput sequencing technologies to convert long RNAs into a cDNA fragments with adaptors attached to one or both ends. Then these fragments are sequenced in a high-throughput manner to obtain short sequences. The resulting reads are aligned to a reference genome or a reference transcript, or assembled ex-novo without the genomic sequence to produce a genome-scale transcription map (Wang et al., 2009). This new method results very useful in the case of non-model organisms for which any genome or transcriptome sequence information is available. Wang et al. (2011) used Illumina's RNA-seq and digital gene expression tag profile (DGE-tag) technologies to screen optimal RNAi targets from *Ostrinia furnalalis* (Guenée) demonstrating that the combination of these two technologies is a rapid, high-throughput, cost less and easy way to select the candidate target genes for RNAi.

One of the biggest limitations to a cost-effective commercial use of this technology is the production of mass quantities of dsRNAs. Tenllado et al. (2004) proved the efficacy of crude preparations of dsRNA in silencing genes of plant viruses using bacterial crude preparations obtained by lysing cell pellets with a French Press avoiding the use of time-consuming protocols to purify nucleic acids. So they anticipated that inducible expression of dsRNA in bacteria could easily be scaled-up as a rapid and cost effective expression system for a large variety of viral dsRNAs. They suggested the use of large fermenters normally used for the commercial production of entomopathogenic nematodes for the mass production of interfering products in agriculture. Moreover they stated that bacterial cell cultures provide a practicable and reproducible approach for mass production of dsRNA because they are easily pelleted, stored and distributed, without losing their interfering properties with plant virus infection.

Concerning the delivery system of dsRNA, the feeding approach instead of the haemocoel injection is more attractive because more feasible in crop protection, that is the target organism should be able to take up the dsRNA autonomously, especially in the case of insects , e.g. by feeding and digestion (Huvenne & Smagghe, 2010). Several studies have proved the possibility of introducing dsRNA in insect through feeding as resumed by Huvenne & Smagghe (2010) and recently confirmed by Pitino *et al.* (2011) who demonstrated that siRNAs can travel from the plant phloem through the aphid *M. persicae* stylet and reach the intestinal tissues triggering the silencing of aphid target genes. Thus the solution for a practical crop protection application is the introduction of dsRNA in the plants that are ingested by insect herbivores. That is possible by *Agrobacterium* infection, by particle bombardment or by virus infection. The first method refers to transgenic RNAi plants that are created, as other genetically modified plants, by means of isolation, amplification and cloning into a plasmid of the target gene. Then the plasmid is allowed to undergo recombination into an *A. tumefaciens* binary vector. Then plants are transformed with *A. tumefaciens* and the vectors replicate to produce ds-RNA for RNAi induction (Johansen & Carrington, 2001). Alternatively bombarding cells with dsRNA can produce transient silencing. On the contrary with Virus Induced Gene Silencing (VIGS) RNA is introduced in the plants by means of modified viruses as RNA silencing trigger (Watson et al., 2005; Wani et al., 2010).

Tenllado et al. (2004) found that a spray made of crude dsRNA was efficient at gene silencing of plant viruses. But even if the dsRNAs were able to stable remain on the leaves for several days conferring resistance to viral infection, this method was recognized not to be cost-effective. Likewise a crude extract of *E. coli* HT115 containing large amounts of dsRNA and applied to maize plants as a spray was able to inhibit Sugarcane mosaic virus (SCMV) infection (Gan et al., 2010). More recently Wang et al. (2011) demonstrated that a direct spray of dsRNA can produce a down-regulation of gene and finally a development delay and death in the Asian corn borer *O. furnalalis* larvae by penetration through the exoskeleton before reaching the haemolimph and producing RNAi effects. This finding opens new perspectives in dsRNA delivery and provides new insight that RNAi targets can be selected from the whole insect instead of gut-specific target after feeding with dsRNA.

Concerning the future of RNAi in crop protection there are good chances for its application in the field as well as interesting commercial implications in particular against viruses and insects. Probably the future of crop protection against insect pests by means of RNAi technology is more related to sap-sucking pests such as aphids, whiteflies, psyllids and leafhoppers. In fact the current expression of Bt insecticidal proteins in plant offers a better degree of protection in field condition against Lepidoptera and Coleoptera, while none of Cry toxins of *B. thuringiensis* is effective on sup-sucking pests (Virla et al., 2010). The good new results obtained with the whitefly *Bemisia tabaci* (Gennadius) (Upadhyay et al., 2011), the aphid *M. persicae* (Pitino et al., 2011) and the brown planthopper *Nilaparvata lugens* Stål (Zha et al., 2011) are promising. Anyhow, recent reports of resistance to Bt toxins expressed in the plants (Tabashnik et al., 2008) will provide more interest for alternative strategies also for Lepidoptera and Coleoptera.

There is also another application of RNAi that could open new opportunities in crop protection. Allowing the specific down-regulation of genes, this technology will permit the identification of new targets for crop protection compounds such as fungicides and insecticides (Busch et al., 2005) for an ever more specific action. With this approach vulnerable targets for new pest-specific insecticides with novel mode of action can be assessed.

Anyhow for a wide practical application of RNAi in crop protection some challenges remain. For instance the long-term effectiveness of the approach will need to be established, in particular it is necessary to investigate how the internal defense mechanism of the insects will respond to long-term exposure to dsRNA.

6. Conclusion

The general trend in crop protection is the progressive reduction of chemicals because of a more consciousness towards human health and environment. Nowadays biotechnology and genomics offer new important possibilities for this purpose. The possibility to transfer genetic information between different organisms allowed to create transgenic plants resistant to different pathogens and pests. The application of this technology to the most economically important crops and the dissemination throughout the world provided a considerable decrease in pesticide use. In parallel genomics help us to understand the precise role of simple sequences and the proteins they encode within the organism opening new perspectives for the production of pesticides with innovative mode of action and for the development of new specific control strategies against plant pathogens and pests such as the RNAi approach. The improvement of the sequencing technology and in particular the advent of the NGS together with the evolution of bioinformatics enable large-scale sequencing projects permitting a new comparative genomics approach and thus it will facilitate the identification of candidate genes to be used in crop protection. The possibility to down-regulate or over express genes in plants or in plant pathogenic organisms is a very powerful tool, within pest management and agriculture in general, to meet farmer and consumer demands.

7. References

Abe, T., Kanaya, S., Kinouchi, M., Ichiba, Y., Kozuki, T. & Ikemura, T. (2003). Informatics for unveiling hidden genome signatures. *Genome Research*, Vol. 13, No. 4, pp. 693–702, ISSN 1088-9051.

Álvarez, A.M. (2004). Integrated approaches for detection of plant pathogenic bacteria and diagnosis of bacterial diseases. *Annual Review of Phytopathology*, Vol. 42, pp. 339-366, ISSN 0066-4286.

Alves, A.P., Lorenzen, M.D., Beeman, R.W., Foster, J.E. & Siegfried, B.D. (2010). RNA interference as a method for target-site screening in the western corn rootworm, *Diabrotica virgifera virgifera*. *Journal of Insect Science*, Vol. 10, No. 162, pp. 1-16, ISSN 1536-2442.

Amdam, G.V., Simoes, Z.L., Guidugli, K.R., Norberg, K. & Omholt, S.W. (2003). Disruption of vitellogenin gene function in adult honeybees by intra-abdominal injection of double-stranded RNA. *BMC biotechnology*, Vol. 3, p. 1, ISSN 1472-6750.

Armstrong, K. (2010). DNA barcoding: a new module in New Zealand's plant biosecurity diagnostic toolbox. *EPPO Bulletin* Vol. 40, No. 1, pp. 91–100, ISSN 1365-2338.

Bakhetia, M., Charlton, W., Atkinson, H.J. & McPherson, M.J. (2005). RNA interference of dual oxidase in the plant nematode *Meloidogyne incognita*. *Molecular Plant-Microbe Interactions*, Vol. 18, No. 10, pp. 1099-1106, ISSN 0894-0282.

Barton, K.A., Whitely, H.R. & Yang, N.-S. (1987). *Bacillus thuringiensis* delta-endotoxin expressed in transgenic *Nicotiana tabacum* provides resistance to lepidopteran insects. *Plant Physiology*, Vol. 85, No. 4, pp. 1103-1109, ISSN 0032-0889.

Bates, S.L., Zhao, J.Z., Roush, R.T. & Shelton, A.M. (2005). Insect resistance management in GM crops: past, present and future. *Nature Biotechnology*, Vol. 23, No.1, pp. 57–62, ISSN 1087-0156.

Baum, J.A., Bogaert, T., Clinton, W., Heck, G.R., Feldmann, P., Ilagan, O., Johnson, S., Plaetinck, G., Munyikwa, T., Pleau, M., Vaughn, T. & Roberts, J. (2007). Control of coleopteran insect pests through RNA interference. *Nature Biotechnology*, Vol. 25, pp. 1322-1326, ISSN 1087-0156.

Bellés, X. (2010). Beyond *Drosophila*: RNAi in vivo and functional genomics in insects. *The Annual Review of Entomology*, Vol. 55, pp.111–128, ISSN 0066-4170.

Bertaccini, A., Paltrinieri, S., Makarova, O., Contaldo N. & Nicolaisen M. (2011). Sharing information and collections on phytoplasmas: from QBOL to QBANK. *Bulletin of Insectology*, Vol. 64 (Supplement), pp. S289-S290, ISSN 1721-8861.

Bertin, S., Picciau, L., Ács Z., Alma, A. & Bosco, D. (2010a). Molecular differentiation of four *Reptalus* species (Hemiptera: Cixiidae). *Bulletin of Entomological Research*, Vol. 100, No. 5, pp. 551–558, ISSN 0007-4853.

Bertin, S., Picciau, L., Ács, Z., Alma, A. & Bosco, D. (2010b), Molecular identification of the *Hyalesthes* species (Hemiptera: Cixiidae) occurring in vineyard agroecosystems. *Annals of Applied Biology*, Vol. 157, No. 3, pp. 435–445, ISSN 0003-4746.

Bonants, P., Groenewald, E., Rasplus, J. Y., Maes, M., De Vos, P., Frey, J., Boonham, N., Nicolaisen, M., Bertacini, A., Robert, V., Barker, I., Kox, L., Ravnikar, M., Tomankova, K., Caffier, D., Li, M., Armstrong, K., Freitas-Astúa, J., Stefani, E., Cubero, J. & Mostert, L. (2010). QBOL: a new EU project focusing on DNA barcoding of Quarantine organisms. *EPPO Bulletin*, Vol. 40, No. 1, pp. 30–33, ISSN 1365-2338.

Boonham, N., Walsh, K., Mumford, R.A. & Barker, I. (2000). The use of multiplex real-time PCR (TaqMan®) for the detection of potato viruses. *EPPO Bulletin*, Vol. 30, No. 3-4, pp. 427–430, ISSN 1365-2338.

Boonham, N., Tomlinson, J. & Mumford, R. (2007). Microarrays for rapid identification of plant viruses. *Annual review of Phytopathology*, Vol. 45, pp. 307–328, ISSN 0066-4286.

Bosher, J.M. & Labouesse, M. (2000). RNA interference: genetic wand and genetic watchdog. *Nature Cell Biology*, Vol. 2, pp. E31–E36, ISSN 1465-7392.

Boulter, D., Gatehouse, A.M.R. & Hilder, V. (1989). Use of cowpea trypsin inhibitor (CpTI) to protect plants against insect predation. *Biotechnology Advances*, Vol. 7, No. 4, pp. 489-497, ISSN 0734-9750.

Bucher, G., Scholten, J. & Klingler, M. (2002). Parental RNAi in *Tribolium* (Coleoptera). *Current Biology*, Vol. 12, pp. R85–R86, ISSN 0960-9822.

Busch, M., Villalba, F., Schulte, T. & Menke, U. (2005). RNAi for discovery of novel crop protection products. *Pflanzenschutz-Nachrichten Bayer*, Vol. 58, No. 1, 34-50, ISSN 0340-1723.

Caterino, M.S., Cho, S. & Sperling F.A.H., (2000). The current state of insect molecular systematics: a thriving Tower of Babel. *Annual Review of Entomology*, Vol. 45, pp. 1–54, ISSN 0066-4170.

Cavalieri, V., Mazzeo, G., Garzia, G.T., Buonocore, E. & Russo, A. (2008). Identification of *Planococcus ficus* and *Planococcus citri* (Hemiptera: Pseudococcidae) by PCR-RFLP of COI gene. *Zootaxa*, Vol. 68, No. May, pp. 65-68, ISSN 1175-5326.

CBOL Plant Working Group (2009). A DNA barcode for land plants. *Proceeding of the National Academy of Sciences USA*, Vol. 106, No. 4, pp. 12794–12797.

Chan, C.K., Hsu, A.L., Tang, S.L., & Halgamuge, S.K. (2008). Using growing self-organising maps to improve the binning process in environmental whole-genome shotgun sequencing. *Journal of Biomedicine and Biotechnology*, Vol. 2008, article ID 513701, ISSN 1110-7243.

Consortium for the Barcode of Life (2009). CBOL approves matK and rbcL as the BARCODE regions for Land Plants. [Cited 15 October 2010.] Available from URL: http://www.barcoding.si.edu/plant_working_group.html.

Contaldo, N., Canel, A., Makarova, O., Paltrinieri, S., Bertaccini, A. & Nicolaisen, M. (2011). Use of a fragment of the tuf gene for phytoplasma 16Sr group/subgroup differentiation. *Bulletin of Insectology*, Vol. 64 (Supplement), pp. S45-S46, ISSN 1721-8861.

Cox, A.J. & Hebert, P.D.N. (2001). Colonization, extinction and phylogeographic patterning in a freshwater crustacean. *Molecular Ecology*, Vol. 10, pp. 371–386, ISSN 0962-1083.

De Leo, F., Volpicella, M., Licciulli, F., Liuni, S., Gallerani, R. & Ceci, L.R. (2002). PLANT-PIs: a database for plant protease inhibitors and their genes. *Nucleic Acids Research*, Vol. 30, pp. 347-348, ISSN 0305-1048.

Díaz-Pendón, J.A. & Ding, S.-W. (2008). Direct and indirect roles of viral suppressors of RNA silencing in pathogenesis. *Annual Review of Phytopathology*, Vol. 46, pp. 303-326, ISSN 0066-4286.

Ding, A., Li, H., Lu, R., Li, F. & Li, W. (2004). RNA silencing: a conserved antiviral immunity of plants and animals. *Science*, Vol. 296, pp. 109–115, ISSN 0036-8075.

Dzitoyeva, S., Dimitrijevic, N. & Manev, H. (2001). Intra-abdominal injection of double-stranded RNA into anesthetized adult *Drosophila* triggers RNA interference in the central nervous system. *Molecular Psychiatry*, Vol. 6, No. 6, pp. 665–670, ISSN 1359-4184.

Escobar, M.A., Civerolo, E.L., Summerfelt, K.R. & Dandekar, A.M. (2001). RNAi-mediated oncogene silencing confers resistance to crown gall tumorigenesis. *Proceeding of the National Academy of Sciences USA*, Vol. 98, N. 23, pp. 13437-13442, ISSN 0027-8424.

Fessehaie, A., De Boer, S.H. & Lévesque, C.A. (2003). An Oligonucleotide array for the identification and differentiation of bacteria pathogenic on potato. *Phytopathology*, Vol. 93, No. 3, pp. 262-269, ISSN 0031-949X.

Fire, A., Xu, S., Montgomery, M., Kostas, S., Driver, S. & Mello, C. (1998). Potent and specific genetic interference by double-stranded RNA in *Caenorhabditis elegans*. *Nature*, Vol. 391, No. 6669, pp. 806–811, ISSN 0028-0836.

Fisk, H.J., & Dandekar, A.M. (1993). The introduction and expression of transgenes in plants. *Scientia Horticulturae*, Vol. 55, No. 1-2, pp. 5-36, ISSN 0304-4238.

Fitzgerald, A., Van Kan, J.A. & Plummer, K.M. (2004). Simultaneous silencing of multiple genes in the apple scab fungus, *Venturia inaequalis*, by expression of RNA with chimeric inverted repeats. *Fungal Genetics and Biology*, Vol. 41, No. 10, pp. 963–971, ISSN 1087-1845.

Florack, D., Allefs, S., Bollen, R., Bosch, D., Visser, B. & Stiekema, W. (1995). Expression of giant silkmoth cecropin B genes in tobacco. *Transgenic Research*, Vol. 4, No. 2, pp. 132-141, ISSN 0962-8819.

Floyd, R., Lima, J., deWaard, JR, Humble, L.R. & Hanner, R.H. (2010). Common goals: incorporating DNA barcoding into international protocols for identification of arthropod pests. *Biological Invasions*, Vol. 12, No. 9, pp. 2947-2954, ISSN 1387-3547.

Foster, S.J., Ashby, A.M. & Fitt B.D.L. (2002). Improved PCR-based assays for pre-symptomatic diagnosis of light leaf spot and determination of mating type of *Pyrenopeziza brassicae* on winter oilseed rape. *European Journal of Plant Pathology*, Vol. 108, No. 4, pp. 379–83, ISSN 0929-1873.

Fraaije, B., Lovll, D.J., Rohel, E.A. & Hollomon, D.W. (1999). Rapid detection and diagnosis of *Septoria tritici* epidemics in wheat using a polymerase chain reaction/Pico green assay. *Journal of Applied Microbiology*, Vol. 86, No. 4, pp. 701-708, ISSN 1364-5072.

Fuchs, M., Provvidenti, R., Slightom, J.L. & Gonsalves, D., (1996). Evaluation of transgenic tomato plants expressing the coat protein gene of cucumber mosaic virus strain WL under field conditions. *Plant Disease*, Vol. 80, No. 3, pp. 270-275, ISSN 0191-2917.

Gan, D.F., Zhang, J., Jiang, H.B., Jiang, T., Zhu, S.W. & Cheng, B.J. (2010). Bacterially expressed dsRNA protects maize against SCMV infection. *Plant Cell Reports*, Vol. 29, No. 11, pp. 1261–1268, ISSN 0721-7714.

Gariepy, T.D., Kuhlmann, U., Gillott, C. & Erlandson, M. (2007). Parasitoids, predators and PCR: the use of diagnostic molecular markers in biological control of Arthropods. *Journal of Applied Entomology*, Vol. 131, No. 4, pp. 225–240, ISSN 0931-2048.

Gatehouse, J.A. (2008). Biotechnological Prospects for engineering insect-resistant plants. *Plant Physiology*, Vol. 146, No. 3, pp. 881-887, ISSN 0032-0889.

Gonsalves, D. (1998). Control of papaya ringspot virus in papaya: a case study. *Annual Review of Phytopathology*, Vol. 36, pp. 415–437, ISSN 0066-4286.

Greenstone, M.H. (2006). Molecular methods for assessing insect parasitism. *Bulletin of Entomological Research*, Vol. 96, No. 1, pp. 1–13, ISSN 0007-4853.

Grover, A. & Gowthaman, R. (2003). Strategies for development of fungus-resistant transgenic plants. *Current Science*, Vol. 84, No. 3, pp. 330–340, 0011-3905.

Haas, B.J. & Zody, M.C. (2010). Advancing RNA-Seq analysis. *Nature Biotechnology*, Vol. 28, No. 5, pp. 421–423, ISSN 1087-0156.

Hammond, T.M. & Keller, N.P.(2005). RNA silencing in *Aspergillus nidulans* is independent of RNA-dependent RNA polymerases. *Genetics*, Vol. 169, No. 2, pp. 607–617, ISSN 0016-6731.

Hebert, P.D.N., Cywinska, A., Ball, S.L. & deWaard, J.R. (2003). Biological identifications through DNA barcodes. *Proceedings of the Royal Society of London Series B-Biological Sciences*, Vol. 270, No. 1512, pp. 313–321, ISSN 0962-8452.

Heckel, D. G. (2003). Genomics in pure and applied entomology. *Annual Review of Entomology*, Vol. 48, pp. 235–260, ISSN 0066-4170.

Henegariu, O., Heerema, N.A., Dlouhy, S.R., Vance, G.H. & Vogt, P.H. (1997). Multiplex PCR: critical parameters and step-by-step protocol. *BioTechniques*, Vol. 23, No. 3, pp. 504-511, ISSN 0736-6205.

Henson, J.M. & French, R. (1993). The polymerase chain reaction and plant disease diagnosis. *Annual Review of Phytopathology*, Vol. 31, pp. 81-109, ISSN 0066-4286.

Hirsch, P.R., Mauchline, T.H., Mendum, T.A. & Kerry, B.R. (2000). Detection of the nematophagous fungus *Verticillium chlamydosporium* in nematode-infested plant roots using PCR. *Mycological Research,* Vol. 104, No. 4, pp. 435–439, ISSN 0953-7562.

Huang, G., Allen, R., Davis, E.L., Baum, T.J. & Hussey, R.S. (2006). Engineering broad root-knot resistance in transgenic plants by RNAi silencing of a conserved and essential root-knot nematode parasitism gene. *Proceedings of the National Academy of Sciences USA,* Vol. 103, No. 5, pp. 14302-14306, ISSN 0027-8424.

Hunter, W., Ellis, J., Vanengelsdorp, D., Hayes, J., Westervelt, D., Glick, E., Williams, M., Sela, I., Maori, E., Pettis, J., Cox-Foster, D., & Paldi, N. (2010). Large-scale field application of RNAi technology reducing Israeli acute paralysis virus disease in honey bees (*Apis mellifera,* Hymenoptera: Apidae). *PLoS Pathogens,* Vol. 6, No. 12, e1001160, ISSN 1553-7366.

Huvenne, H. & Smagghe, G. (2010). Mechanisms of dsRNA uptake in insects and potential of RNAi for pest control: A review. *Journal of Insect Physiology,* Vol. 56, No. 3, pp. 227-235, ISSN 0022-1910.

Jacobs, B.L. & Langland, J.O. (1996). When two strands are better than one: the mediators and modulators of the cellular responses to double stranded RNA. *Virology,* Vol. 219, No. 2, pp. 339–349, ISSN 0042-6822.

Jaubert-Possamai, S., Le Trionnaire, G., Bonhomme, J., Christophides, G. K., Rispe, C., & Tagu, D. (2007). Gene knockdown by RNAi in the pea aphid *Acyrthosiphon pisum. BMC Biotechnology,* Vol. 7, No. 1, p. 63, ISSN 1472-6750.

Jiménez-Gasco, M.M., Milgroom, M.G., Jiménez-Díaz, R.M. (2002). Gene genealogies support *Fusarium oxysporum* f. sp. *ciceris* as a monophyletic group. *Plant Pathology,* Vol. 51, No. 1, pp. 72–77, ISSN 0032-0862.

Jinbo, U., Kato T. & Ito M. (2011). Current progress in DNA barcoding and future implications for entomology. *Entomological Science,* Vol. 14, No. 2, pp. 107–124, ISSN 1479-8298.

Johansen, L.K., & Carrington, J.C. (2001). Silencing on the spot: Induction and suppression of RNA silencing in the *Agrobacterium*-mediated transient expression system. *Plant Physiology,* Vol. 126, No. 3, pp. 930–938, ISSN 0032-0889.

Jongedijk, E., de Schutter, A.A.J.M., Stolte, T., van den Elzen, P.J.M. & Cornelissen, B.J.C., (1992). Increased resistance to potato virus X and preservation of cultivar properties in transgenic potato under field conditions. *BioTechnology (N.Y.),* Vol. 10, No. 4, pp. 422-429, ISSN 0733-222X.

Kadotani, N., Nakayashiki, H., Tosa, Y. & Mayama, S. (2003). RNA silencing in the phytopathogenic fungus *Magnaporthe oryzae. Molecular Plant–Microbe Interactions,* Vol. 16, No. 8, pp. 769-776, ISSN 0894-0282.

Karakaş, M. (2010). RNA interference in plant parasitic nematodes. *African Journal of Biotechnology,* Vol. 7, No. 15, pp. 2530-2534, ISSN 1684-5315.

Ketting, R.F. & Plasterk, R.H.A. (2000). A genetic link between cosupression and RNA interference in *C. elegans. Nature,* Vol. 404, No. 6775, pp. 296–298, ISSN 0028-0836.

Kumar, M. & Carmichael, G.G. (1998). Antisense RNA: function and fate of duplex RNA in cells of higher eukaryotes. *Microbiology and Molecular Biology Reviews,* Vol. 62, No. 4, pp. 1415-1434, ISSN 1092-2172.

Lévesque, C.A., Harlton, C.E. & de Cock, A.W. (1998). Identification of some oomycetes by reverse dot blot hybridization. *Phytopathology*, Vol. 88, No. 3, pp. 213-222, ISSN 0031-949X.

Li, H., Xiang, W. & Ding, S.W. (2002). Induction and suppression of RNA silencing by an animal virus. *Science*, Vol. 296, No. 5571, pp. 1319-1321, ISSN 0036-8075.

Li, F. & Ding, S.W. (2006). Virus counter defense: diverse strategies for evading the RNA silencing immunity. *Annual Review of Microbiology*, Vol. 60, pp. 503–531, ISSN 0066-4227.

Lievens, B., Brouwer, M., Vanachter, A.C.R.C., Levesque C.A., Cammue B.P.A. & Thomma B.P.H.J. (2003). Design and development of a DNA array for rapid detection and identification of multiple tomato vascular wilt pathogens. *FEMS Microbiology Letters*, Vol. 223, No. 1, pp. 113-122, ISSN 1574-6968.

Lievens, B. & Thomma, B.P.H.J. (2005). Recent developments in pathogen detection arrays: Implications for fungal plant pathogens and use in practice. *Phytopathology*, Vol. 95, No. 12, pp. 1374-1380, ISSN 0031-949X.

Lilley, C.J., Bakhetia, M., Charlton, W.L. & Urwin, P.E. (2007). Recent progress in the development of RNA interference for plant parasitic nematodes. *Molecular Plant Pathology*, Vol. 8, No. 5, pp. 701-711, ISSN 1364-3703.

Liu, H., Cottrell, T.R., Pierini, L.M., Goldman, W.E. & Doering, T.L. (2002). RNA interference in the pathogenic fungus *Cryptococcus neoformans*. *Genetics*, Vol. 160, No. 2, pp. 463–470, ISSN 1471-0056.

López, M.L., Bertolini, E., Olmos, A., Caruso, P., Goris, M.T., Llop, P., Penyalver, R. & Cambra, M. (2003). Innovative tools for detection of plant pathogenic viruses and bacteria. *International Microbiology*, Vol. 6, No. 4, pp. 233-243, ISSN 1139-6709.

Louws, F.J., Rademaker, J.L.W. & de Bruijn, F.J. (1999). The three Ds of PCR-based genomic analysis of phytobacteria: diversity, detection, and disease diagnosis. *Annual Review of Phytopathology*, Vol. 37, pp. 81–125, ISSN 0066-4286.

Malnoe, P., Farinelli, L., Collet, G.F. & Reust, W. (1994). Small-scale field tests with transgenic potato, cv. Bintje, to test resistance to primary and secondary infections with potato virus Y. *Plant Molecular Biology*, Vol. 25, No. 6, pp. 963-975, ISSN 0167-4412.

McCartney, H.A., Foster, S.J., Fraaije, B.A. & Ward, E. (2003). Molecular diagnostics for fungal plant pathogens. *Pest Management Science*, Vol. 59, No. 2, pp. 129-142, ISSN 1526-4998.

Michiels, K., Van Damme, E.J. & Smagghe, G. (2010). Plant-insect interactions: what can we learn from plant lectins? *Archives of Insect Biochemistry and Physiology*, Vol. 73, No. 4, pp. 193–212, ISSN 1520-6327.

Misquitta, L. & Paterson, B.M. (1999). Targeted disruption of gene function in *Drosophila* by RNA interference (RNA-i): A role for nautilus in embryonic somatic muscle formation. *Proceedings of Natural Academic of Sciences USA*, Vol. 96, No. 4, pp. 1451–1456, ISSN 0027-8424.

Mito, T., Nakamura, T., Bando, T., Ohuchi, H. & Noji, S. (2011). The advent of RNA interference in Entomology. *Entomological Science*, Vol. 14, No. 1, pp. 1–8, ISSN 1479-8298.

Moritz, C. & Cicero, C. (2004). DNA barcoding: promise and pitfalls. *PLoS Biology*, Vol. 2, No. 10, e354, ISSN 1544-9173.

Mouyna, I., Henry, C., Doering, T.L. & Latgé, J.P. (2004). Gene silencing with RNA interference in the human pathogenic fungus *Aspergillus fumigatus*. *FEMS Microbiology Letters*, Vol. 237, No. 2, pp. 317–324, ISSN 1574-6968.

Mumford, R., Boonham, N., Tomlinson, J. & Barker, I. (2006). Advances in molecular phytodiagnostics - new solutions for old problems. *European Journal of Plant Pathology*, Vol. 116, No. 1, pp. 1–19, ISSN 0929-1873.

Murdock, L.L. & Shade, R.E. (2002). Lectins and protease inhibitors as plant defenses against insects. *Journal of Agricultural and Food Chemistry*, Vol. 50, No. 22, pp. 6605-6611, ISSN 0021-8561.

Mutti, N.S., Park, Y., Reese, J.C. & Reeck, G.R. (2006). RNAi knockout of a salivary transcript leading to lethality in the pea aphid, *Acyrthosiphon pisum*. *Journal of Insect Science*, Vol. 6, No. 38, pp. 1-7, ISSN 1536-2442.

Napoli, C., Lemiex, C. & Jorgensen R. (1990). Introduction of a chimeric chalcone synthase gene into petunia results in reversible co-suppression of homologous genes in trans. *The Plant Cell*, Vol. 2, No. 4, pp. 279-289, ISSN 1040-4651.

Neumann, G., Follett, P.A., Hollingsworth, R.G. & de Leon, J.H. (2010). High host specificity in *Encarsia diaspidicola* (Hymenoptera: Aphelinidae), a biological control candidate against the white peach scale in Hawaii. *Biological Control*, Vol. 54, No. 2, pp. 107–113, ISSN 1049-9644.

O'Donnell, K., Kistler, H.C., Cigelink, E. & Ploetz, R.C. (1998). Multiple evolutionary origins of the fungus causing Panama disease of banana: Concordant evidence from nuclear and mitochondrial gene genealogies. *Proceedings of the National Academy of Sciences USA*, Vol. 95, No. 5, pp. 2044-2049, ISSN 0027-8424.

Pitino, M., Coleman, A.D., Maffei, M.E., Ridout, C.J. & Hogenhout, S.A. (2011). Silencing of aphid genes by dsRNA feeding from plants. *PLoS ONE*, Vol. 6, No. 10, e25709, ISSN 1932-6203.

Powel A.P., Stark, D.M., Sanders, P.R. & Beachy, R.N. (1986). Protection against tobacco mosaic virus in transgenic plants that express tobacco mosaic virus antisense RNA. *Proceedings of the National Academy of Sciences USA*, Vol. 86, No. 18, pp. 6949-6952, ISSN 0027-8424.

Presting, G.G., Smith, O.P. & Brown, C.R. (1995). Resistance to potato leafroll virus in potato plants transformed with the coat protein gene or with vector control constructs. *Phytopathology*, Vol. 85, No. 4, pp. 436-442, ISSN 0031-949X.

Price, D.R. & Gatehouse, J.A. (2008). RNAi-mediated crop protection against insects. *Trends in Biotechnology*, Vol. 26, No. 7, pp. 393–400, ISSN 0167-7799.

Prins, M., De Haan, P., Luyten, R., Van Veller, M., Van Grinsven, M.Q.J.M. & Goldbach, R. (1995). Broad resistance to tospoviruses in transgenic plants by expressing three tospoviral nucleoprotein gene sequences. *Molecular Plant–Microbe Interactions*, Vol. 8, No. 1, pp. 85–91, ISSN 0894-0282.

Prins, M., Laimler, M., Noris, E., Schubert, J., Wassenegger, M., & Tepfer, M. (2008). Strategies for antiviral resistance in transgenic plants. *Molecular Plant Pathology*, Vol. 9, No. 1, pp. 73-83, ISSN 1364-3703.

Quan, P.-L., Briese, T., Palacios, G., & Lipkin, W.I. (2008). Rapid sequence-based diagnosis of viral infection. *Antiviral Research*, Vol. 79, No. 1, pp. 1-5, ISSN 0166-3542.

Rao, K.V., Rathore, K.S., Hodges, T.K., Fu, X., Stoger, E., Sudhakar, D., Williams, S., Christou, P., Bharathi, M., Bown, D.P., Powell, K.S., Spence, J., Gatehouse, A.M. &

Gatehouse, J.A. (1998). Expression of snowdrop lectin (GNA) in transgenic rice plants confers resistance to rice brown planthopper. *The Plant Journal*, Vol. 15, No. 4, pp. 469-477, ISSN 1365-313X.

Romano, N. & Macino, G. (1992) Quelling: transient inactivation of gene expression in *Neurospora crassa* by transformation with homologous sequences. *Molecular Microbiology*, Vol. 6, No. 22, pp. 3343-3353, ISSN 1365-2958.

Rosso, M.N., Dubrana, M.P., Cimbolini, N., Jaubert, S. & Abad, P. (2005). Application of RNA interference to root-knot nematode genes encoding esophageal gland proteins. *Molecular Plant-Microbe Interactions*, Vol. 18, No. 7, pp. 615-620, ISSN 0894-0282.

Rosso, M.N., Jones, J.T. & Abad, P. (2009). RNAi and functional genomics in plant parasitic nematodes. *Annual Review of Phytopathology*, Vol. 47, pp. 207-232, ISSN 0066-4286.

Saccaggi, D.L., Krüger, K. & Pietersen, G. (2008). A multiplex PCR assay for the simultaneous identification of three mealybug species (Hemiptera: Pseudococcidae). *Bulletin of Entomological Research*, Vol. 98, No. 1, pp. 27-33, ISSN 0007-4853.

Sanders, P.R., Sammons, B., Kaniewski, W., Haley, L., Layton, J., Lavallee, B.J., Delannay, X. & Tumer, N.E. (1992). Field resistance of transgenic tomatoes expressing the tobacco mosaic virus or tomato mosaic virus coat protein genes. *Phytopathology*, Vol. 82, No. 6, pp. 683-690, ISSN 0031-949X.

Sanford, J.C. & Johnson, S.A. (1985). The concept of parasite-derived resistance: deriving resistance genes from the parasites own genome. *Journal of Theoretical Biology*, Vol. 115, No. 3, pp. 395-405, ISSN 0022-5193.

Schaad, N.W. & Frederick, R.D. (2002). Real-time PCR and its application for rapid plant disease diagnostics. *Canadian Journal of Plant Pathology*, Vol. 24, No. 3, pp. 250-258, ISSN 0706-0661.

Schaad, N.W., Frederick, R.D., Shaw, J., Schneider, W.L., Hickson, R., Petrillo, M. & Luster, D.G. (2003). Advances in molecular-based diagnostics in meeting crop biosecurity and phytosanitary issues. *Annual Review of Phytopathology*, Vol. 41, pp. 305-324, ISSN 0066-4286.

Schmid A., Schindelholz, B. & Zinn, K. (2002). Combinatorial RNAi: a method for evaluating the functions of gene families in *Drosophila*. *Trends in Neurosciences*, Vol. 25, No. 2, pp. 71-74, ISSN 0166-2236.

Seifert, K.A. (2009). Progress towards DNA barcoding of fungi. *Molecular Ecology Resources*, Vol. 9, No. 1, pp. 83–89, ISSN 1755-098X.

Stoeckle, M. (2003). Taxonomy, DNA and the bar code of life. *BioScience*, Vol. 53, No. 1, pp. 2–3, ISSN 0006-3568.

Strittmatter, G. & Wegener. D. (1993). Genetic engineering of disease and pest resistance in plants: present state of the art. *Zeitschrift für Naturforschung C, a journal of Biosciences*, Vol. 48, No. 9-10, pp. 673–688, ISSN 0939-5075.

Studholme, D.J., Glover R.H. & Boonham, N. (2011). Application of High-Throughput DNA Sequencing in Phytopathology. *Annual Review of Phytopathology*, Vol. 49, pp. 87-105, ISSN 0066-4286.

Symondson, W.O.C. (2002). Molecular identification of prey in predator diets. *Molecular Ecology*, Vol. 11, No. 4, pp. 627–641, ISSN 0962-1083.

Tabara, H., Grishok, A. & Mello, C.C. (1998). RNAi in *C. elegans*: soaking in the genome sequence. *Science*, Vol. 282, No. 5388, pp. 430–431, ISSN 0036-8075.

Tabashnik, B.E., Gassmann, A.J., Crowder, D.W. & Carriere, Y. (2008). Insect resistance to Bt crops: evidences versus theory. *Nature Biotechnology*, Vol. 26, No. 2, pp. 199–202, ISSN 1087-0156.

Tavernarakis, N., Wang, S.L., Dorovkov, M., Ryazanov, A. & Driscoll, M. (2000). Heritable and inducible genetic interference by double-stranded RNA encoded by transgenes. *Nature Genetics*, Vol. 24, No. 2, pp. 180–183, ISSN 1061-4036.

Tedeschi, R. & Nardi, F. (2010). DNA-based discrimination and frequency of phytoplasma infection in the two hawthorn-feeding species, *Cacopsylla melanoneura* and *Cacopsylla affinis*, in northwestern Italy. *Bulletin of Entomological Research*, Vol. 100, No. 6, pp. 741-747, ISSN 0007-4853.

Tenllado, F., Llave, C. & Diaz-Ruiz, J.R. (2004). RNA interference as a new biotechnological tool for the control of virus diseases in plants. *Virus Research*, Vol. 102, No. 2, pp. 85–96, ISSN 0168-1702.

Timmons, L. & Fire, A. (1998). Specific interference by ingested dsRNA. *Nature*, Vol. 395, No. 6705, p. 854, ISSN 0028-0836.

Timmons, L., Court, D.L. & Fire, A. (2001). Ingestion of bacterially expressed dsRNAs can produce specific and potent genetic interference in *Caenorhabditis elegans*. *Gene*, Vol. 263, No. 1-2, pp. 103–112, ISSN 0378-1119.

Tomoyasu, Y. & Denell, R.E. (2004). Larval RNAi in *Tribolium* (Coleoptera) for analyzing adult development. *Development Genes and Evolution*, Vol. 214, No. 11, pp. 575-578, ISSN 0949-944X.

Tomoyasu, Y., Miller, S.C., Tomita, S., Schoppmeier, M., Grossmann, D. & Bucher, G. (2008) Exploring systemic RNA interference in insects: a genome-wide survey for RNAi genes in *Tribolium*. *Genome Biology*, Vol. 9, No. 1, R10, ISSN 1465-6906.

Tooley, P.W., Carras, M.M. & Falkenstein, K.F. (1996). Relationship among group IV *Phytophthora* species inferred by the restriction analysis of the ITS2 region. *Journal of Phytopathology*, Vol. 144, No. 7-8, pp. 363–369, ISSN 0931-1785.

Turner, C.T., Davy, M.W., MacDiarmid, R.M., Plummer, K.M., Birch, N.P. & Newcomb, R.D. (2006). RNA interference in the light brown apple moth, *Epiphyas postvittana* (Walker) induced by double-stranded RNA feeding. *Insect Molecular Biology*, Vol. 15, No. 3, pp. 383-391, ISSN 1365-2583.

Uehara, T., Kushida, A. & Momota, Y. (1999). Rapid and sensitive identification of Pratylenchus spp. using reverse dot blot hybridization. *Nematology*, Vol. 1, No. 5, pp. 549-555, ISSN 1388-5545.

Unruh, T.R. & Woolley, J.B. (1999). Molecular methods in classical biological control, In: *Handbook of Biological Control*, Bellows T.S. & Fisher T.W. (eds), pp. 57–85. Academic Press, ISBN 0-12-257305-6, New York, USA .

Upadhyay, S.K., Chandrashekar, K., Thakur, N., Verma, P.C., Borgio, J.F., Singh, P.K. & Tuli R. (2011). RNA interference for the control of whiteflies (*Bemisia tabaci*) by oral route. *Journal of Biosciences*, Vol. 36, No. 1, pp. 153-161, ISSN 0250-5991.

Urwin, P.E., Lilley, C.J. & Atkinson, H.J. (2002). Ingestion of double-stranded RNA by pre parasitic juvenile cyst nematodes leads to RNA interference. *Molecular Plant-Microbe Interactions*, Vol. 15, No. 8, pp. 747-752, ISSN 0894-0282.

Vaeck, M., Reynaerts, A., Hofte, H., Jansens, S., De Beuckeler, M., Dean, C., Zabeau, M., Van Montagu, M. & Leemans, J. (1987). Transgenic plants protected from insect attack. *Nature*, Vol. 328, No. 6125, pp. 33-37, ISSN 0028-0836.

Valentini, A., Miquel, C., Nawaz, M.A., Bellemain, E., Coissac, E., Pompanon, F., Nascetti, G. Wincker, P., Swenson, J.E. & Taberlet, P. (2008). New perspectives in diet analysis based on DNA barcoding and parallel pyrosequencing: the trnL approach. *Molecular Ecology*, Vol. 9, No. 1, pp. 51-60, ISSN 0962-1083.

Valentini, A., Pompanon, F. & Taberlet, P. (2009). DNA Barcoding for ecologists. *Trends in Ecology and Evolution*, Vol. 24, No. 2, pp. 110-117, ISSN 0169-5347.

Vijayan, K. & Tsou, C.H. (2010). DNA barcoding in plants: taxonomy in a new perspective. *Current Science*, Vol. 99, No. 11, pp. 1530-1541, ISSN 0011-3905.

Virla, E.G., Casuso, M. & Frias, E.A. (2010). A preliminary study on the effects of a transgenic corn event on the non target pest *Dalbulus maidis* (Hemitera: Cicadeliidae). *Crop Protection*, Vol. 29, No. 6, pp. 635–638, ISSN 0261-2194.

Wallace, M.M. & Covert S.F. (2000). Molecular mating type assay for *Fusarium circinatum*. *Applied Environmental Microbiology*, Vol. 66, No. 12, pp. 5506-5508, ISSN 0099-2240.

Wang, Z., Gerstein, M. & Snyder, M. (2009). RNA-Seq: a revolutionary tool for transcriptomics. *Nature Reviews Genetics*, Vol. 10, No. 1, pp. 57–63, ISSN 1471-0056.

Wang, Y., Zhang, H., Li, H. & Miao, X. (2011). Second-generation sequencing supply an effective way to screen RNAi targets in large scale for potential application in pest insect control. *PLoS ONE*, Vol. 6, No. 4, e18644, ISSN 1932-6203.

Wani, S. & Sanghera, G. (2010). Genetic Engineering for Viral Disease Management in Plants. *Notulae Scientia Biologicae*, Vol. 2, No. 1, pp. 20-28, ISSN 2067-3205.

Wani, S.H., Sanghera, G.S. & Singh, N.B. (2010). Biotechnology and Plant Disease Control-Role of RNA Interference. *American Journal of Plant Sciences*, Vol. 1, No. 2, pp. 55-68, ISSN 2158-2742.

Wares, J.P. & Cunningham, C.W. (2001). Phylogeography and historical ecology of the North Atlantic intertidal. *Evolution*, Vol. 55, No. 12, pp. 2455–2469, ISSN 0014-3820.

Waterhouse, P.M., Wang, M.-B. & Lough, T. (2001). Gene silencing as an adaptive defence against viruses. *Nature*, Vol. 411, No. 6839, pp. 834–842, ISSN 0028-0836.

Watson, J.M., Fusaro, A.F., Wang, M. & Waterhouse, P.M. (2005). RNA silencing platforms in plants. *FEBS Letters*, Vol. 579, No. 26, pp. 5982–5987, ISSN 0014-5793.

Weiland, J.J. & Sundsbak, J.L. (2000). Differentiation and detection of sugar beet fungal pathogens using PCR amplification of actin coding sequences and the ITS region of the rRNA gene. *Plant Disease*, Vol. 84, No. 4, pp. 475-482, ISSN 0191-2917.

Weller, S.A., Elphinstone, J.G., Smith, N., Boonham, N. & Stead, D.E. (2000). Detection of *Ralstonia solanacearum* strains using a quantitative, multiplex, real-time, fluorogenic PCR (TaqMan) assay. *Applied and Environmental Microbiology*, Vol. 66, No. 7, pp. 2853-2858, ISSN 0099-2240.

White, T.J., Bruns, T., Lee, S. & Taylor, J. (1990). Amplification and direct sequencing of fungal ribosomal RNA genes for phylogenetics, In: PCR *Protocols: a guide to methods and applications*. Innis M.A., Gelfand D.H., Sninsky J.J., White T.J., eds., pp. 315–322, Academic Press, ISBN 0-12-372181-4, New York, USA.

Whyard, S., Singh, A.D. & Wong, S. (2009). Ingested double-stranded RNAs can act as species-specific insecticides. *Insect Biochemistry and Molecular Biology*, Vol. 39, No. 11, pp. 824-832, ISSN 0965-1748.

Wolfe, K.H., Li, W.H. & Sharp, P.M. (1987). Rates of nucleotide substitution vary greatly among plant mitochondrial, chloroplast and nuclear DNAs. *Proceedings of the National Academy of Sciences USA,* Vol. 84, No. 24, pp. 9054–9058, ISSN 0027-8424.

Zha, W., Peng, X., Chen, R., Du, B., Zhu, L. & He, G. (2011). Knockdown of midgut genes by dsRNA-transgenic plant-mediated RNA interference in the Hemipteran insect *Nilaparvata lugens. PLoS ONE,* Vol. 6, No. 5, e20504, ISSN 1932-6203.

Zhang, X., Sato, S., Ye, X., Dorrance, A.E., Morris, T.J., Clemente, T.E. & Qu, F. (2011). Robust RNAi-based resistance to mixed infection of three viruses in soybean plants expressing separate short hairpins from a single transgene. *Phytopathology,* Vol. 101, No. 11, pp. 1264-1269, ISSN 0031-949X.

Zhao, J., Cao, J., Li, Y., Collins, H., Roush, R., Earle, E. & Shelton, A. (2003). Transgenic plants expressing two *Bacillus thuringiensis* toxins delay insect resistance evolution. *Nature Biotechnology,* Vol. 21, No. 12, pp. 1493-1497, ISSN 1087-0156.

4

The Input of DNA Sequences to Animal Systematics: Rodents as Study Cases

Laurent Granjon[1] and Claudine Montgelard[2]
[1]Institut de Recherche Pour le Développement (IRD),
CBGP(UMRIRD/INRA/CIRAD/MontpellierSupAgro), Campus de Bel-Air, Dakar,
[2]Biogéographie et Ecologie des Vertébrés (EPHE),
Centre d'Ecologie Fonctionnelle et Evolutive (UMR 5175 CNRS), Montpellier, Cedex 5
[1]Senegal
[2]France

1. Introduction

1.1 General context

The advent of molecular techniques, and especially the possibility to get DNA sequences that can then be compared between individuals of any taxon using more and more powerful algorithms of analysis, has represented a kind of revolution in systematics (Lecointre et al., 2006; Giribet et al., 2007). As in other groups, mammalian systematics was for long only based on morphological and anatomical characters. A brief history of Mammal taxonomy has been provided by Wilson & Reeder (2005: xxiii), starting by early works of Trouessart (1898-99, 1904-5) and ending by the compilation by Mc Kenna & Bell (1997). Since the latter, a huge quantity of data including a significant proportion of molecular ones has accumulated, that have greatly improved our view of the relationships between main mammalian groups (Springer et al., 2004), and increased the number of individual species in each of them (Wilson & Reeder, 2005). Rodents represent the most diverse order of Mammals, comprising around 40% of both generic and specific mammalian diversity (Wilson & Reeder, 2005). In one of the first significant contribution to the study of their classification, Tullberg (1899) subdivided them into two suborders, the Sciurognathi and Hystricognathi (see Hautier et al., 2011), based on morpho-anatomical characteristics of their skull. Several decades of research have led to significant advances in our understanding of rodent evolutionary systematics that were synthesized in Luckett & Hartenberger (1985) major contribution. Again, arrangement of the diversity at the various taxonomic levels, as well as species characterization and delimitation in this group, have greatly improved with the growing use of nuclear and mitochondrial gene sequence data since the 80's, and more importantly the 90's (Catzeflis et al., 1992; Carleton & Musser, 2005).

This review aims at showing how DNA sequences have re-boosted rodent systematics, bringing new data in support or in contradiction to traditional ones, but finally leading to a much better supported classification of this order, via a more accurate characterization and delimitation of its constituent subgroups, down to the species level. It is not intended to be a plea for what has been critically quoted as "the molecularisation of taxonomy" by Lee (2004), but rather to acknowledge the progress in systematics (using rodents as model cases) that DNA sequence data have brought in an integrative context. Indeed, it is clearly within this philosophy of "delimit[ing] the units of life's diversity from multiple and complementary perspectives" (*sensu* Dayrat, 2005: 407) that spectacular advances have been made in recent years in the characterization of taxa and the assessment of their phylogenetic relationships. Saying that implicitly implies that phylogeny should serve as basis for taxonomy, a principle which is underlying most of the current works and findings in these disciplines and which will not be discussed further here.

After some methodological considerations, this chapter chapter will follow a top-down organization in taxonomic ranking, from higher taxa to species, and will be based on numerous examples taken in various rodent groups, and studied with a variety of DNA sequence data.

1.2 Advantages, drawbacks and progress of molecular data

Evolutionary relationships between organisms are generally expressed in phylogenetic trees which then may serve as basis for classification and systematics. A phylogenetic tree is termed fully resolved when only two descending branches are issuing from each node. Such a node is a dichotomy and it represents the speciation event that occurred between two taxa from their hypothetical most recent common ancestor. If there are more than two descending branches, the node is a polytomy or a multifurcation. In this case, the interpretation is more difficult because it can actually mean that an ancestral taxon has split into several descendant taxa but it can also signify that the data used do not resolve the dichotomous branching orders between taxa, and the node is thus unresolved. The concepts of soft and hard polytomies have been introduced (Maddison, 1989; Poe & Chubb, 2004; Walsh et al., 1999) to distinguish between phylogenetic irresolutions due to inadequate data and/or methods, and rapid radiations. The term radiation is used to describe a phase of "divergent evolution of numerous related lineages within a relatively short time" (Futuyma, 1998 : 117; Figure 1A). In the case of soft polytomies, increasing the number of characters analyzed should bring more information and thus solve dichotomous nodes previously masked. If the increase of the number of characters used does not help to decipher phylogenetic relationships, then hard dichotomies should be suspected. However, the question will remain as to know if enough data have been added to resolve the multifurcation. Different molecular studies have shown that an increase in the number of characters combined to the analysis of new genes (in general nuclear exons or introns) have effectively helped to disentangle polytomies that were reputed hard or at least very difficult to solve. Among them are the relationships among the different orders of mammals (Madsen et al., 2001; Murphy et al., 2001), the different families of birds of the super-order Neoaves (Ericson et al., 2006), or the different families of rodents (see paragraphs 2.2.2 and 2.3.1). In these different examples, molecular studies led to some inferences about lineage relationships that were never previously hypothesized from morphological characters.

Advantages of molecular data over traditional morphological characters have rapidly been identified: they represent objective characters whose number can be increased nearly to infinity, and they allow comparisons between phylogenetically distant organisms even when there is no comparable morphological character between the corresponding taxa. However, more time has been needed to realize that homoplasy (convergence and reversal of characters) was a major issue that could blur the phylogenetic signal and thus lead to erroneous groupings (see paragraph 2.1.2). This is particularly problematic in the case of rapid radiations when phylogenetic signal has to be recorded during short time intervals (internodes) whereas this signal might be eroded during the long time span of individual lineage evolution (Whitfield & Lockhard, 2007; Figure 1). The art of phylogeny consists in establishing robust relationships, especially in such circumstances, while avoiding biases due to homoplasy.

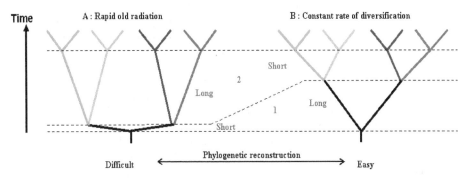

1: Divergence time between basal nodes; 2: Divergence time between terminal nodes

Fig. 1. Some phylogenetic reconstructions are harder to solve than others (modified from Whitfield & Lockhard, 2007).

Various tools are nevertheless available to avoid the misleading effects of homoplasy. The first one is *the choice of the gene*. The molecular era started with the sequencing of mitochondrial genes (cytochrome b, 12S rRNA, control region), mainly because they were easily amplified (several copies by cell as compared to only one copy of nuclear DNA). Several years passed before it was realized that the mitochondrial genome is not adapted to address questions related to ancient divergence events, such as for example the diversification of mammalian orders (Honeycutt & Adkins, 1993) or the radiation of Rodentia (see paragraph 2.1.2). The reason is that the rate of evolution of mitochondrial genes is generally much higher than the nuclear ones. The consequence is that homoplasy accumulates more rapidly (especially at third position of coding genes), thus hiding the phylogenetic signal. The advent of nuclear gene sequencing allowed to resolve many polytomies because of a lesser homoplastic signal. This does not mean that mitochondrial genes have to be banished but rather that it is necessary to evaluate what gene(s) will be the most adapted to the phylogenetic questions asked. In short, rapid-evolving gene will fit for recent phylogenetic separation and slow-evolving genes for ancient radiation.

The choice of the reconstruction method is also of prime importance to avoid artificial groupings. The first molecular trees were produced using the same reconstruction methods as for morphological characters, namely cladistic analysis. However, as soon as 1978, Felsenstein drew the phylogeneticist's attention about artifacts of reconstruction, among which the famous "long-branch attraction", that is the grouping of unrelated taxa on the basis of convergent mutations due to high rates of evolution. The advent of probabilistic methods (Maximum likelihood and Bayesian inference) based on explicit models of sequence evolution greatly improved tree reconstructions (see Felsenstein, 2004). These models include different parameters, such as base frequency, type of substitution or substitution-rate heterogeneity. More parameters will likely be included in the future as their influence on sequence evolution is discovered.

Consequently, *the choice of the model of sequence evolution* is also a decisive stage to cope with homoplasy, as the main objective is to correct effects of hidden multiple substitutions leading to underestimation of the real distance between taxa. Estimation of the model of sequence evolution and associated parameters the best adapted to the dataset is now the first step when starting a phylogenetic analysis. These models not only concern the whole studied genes but also partitions of them according to their functional structure (or any other kind of partition). For example, if the analysis is based on several coding genes, it can be more fruitful to partition the genes according to the three coding positions than considering the whole genes separately. The main inconvenience is that most softwares do not implement these partitioned models yet, but this is a promising way to handle sequence homoplasy. More recently, mixture models have been developed that are based on site-specific patterns of substitution (see Philippe et al., 2005) allowing a better description of the substitution process, thus a reduction of systematic errors in tree reconstructions.

As the number of possible trees is nearly infinite (more than 30 millions of rooted trees for 10 taxa), *the test of node robustness* quickly appeared as a necessity. The most widespread method is the bootstrap (Felsenstein, 1985) which rebuilds phylogenetic trees from artificial matrices created after character resampling (drawing of n sites among N). The number of times (in percentage) each majority grouping has been observed among the reconstructed trees is then indicated at nodes. Another method consists in comparing the best tree to one or several alternative hypotheses. This method arose from the development of probabilistic analyses that attribute a likelihood value (conditional probability) to each tree. Then, it is possible to test if the difference in loglikelihood between two trees is statistically significant or not, and thus to answer the question: is the alternative tree significantly worse than the null hypothesis (in general the optimal phylogenetic tree)?

The advent of phylogenomics and Next Generation Sequencing (NGS) combined with sophisticated probabilistic methods of tree-building using complex models of sequence evolution, implicitly led to the belief that the era of biased and artefactual reconstructions was over. However, increasing the number of nucleotides does not necessarily resolve phylogenetic incongruence but, on the contrary, may lead to incorrect although well-supported trees (Delsuc et al., 2005; Jeffroy et al., 2006). So, it is still important to test the quality of the dataset by estimating its homoplasy content. Some studies recommend the removal of factors of inconsistency, such as fast-evolving positions, or positions with compositional biases (see Philippe et al., 2005). Given the increasing number of available nucleotides, removing some parts of the sequences appears feasible because theoretically enough informative positions should remain to ensure consistency of the phylogenetic signal. It is thus likely that in the future, combining an increasing number of molecular characters through NGS with powerful phylogenetic analyses will conduct to reduce again the number of polytomies, if any are left!

2. Higher levels of Rodentia systematics

2.1 Systematic position of the order Rodentia among mammals and the question of rodent monophyly

2.1.1 The consensual pre-molecular era

From a morphological point of view, the main character defining Rodentia is the presence of a unique large evergrowing chisel-like incisor by half-jaw. This is however not the only distinctive character as no less than six other characters were listed in Hartenberger (1985: 10). Moreover, all these features are clearly derived eutherian characters (synapomorphies), thus making monophyly of Rodentia strongly supported. By contrast to other mammalian orders (for example Artiodactyla or Carnivora) that were delimited with more difficulty (see Simpson, 1945), Rodentia appeared morphologically well characterized. This concerns not only living rodents but more surprisingly also fossils. This was clearly stated by Simpson (1945: 198): "...the order is exceptionally clear cut. There is not, even among fossils, any question as to whether a given animal is or is not a rodent, however doubtful its position in the order may be." So, before the nineties, no scientist contested the monophyly of the group and the main questions regarding Rodentia concerned higher (identification of its sister taxon) or lower (relationships between families; see paragraph 2.2) taxonomic levels.

The question of the systematic position of Rodentia among the other mammalian orders, although more discussed than rodent monophyly, did not raise much contradictory debates. In fact only two hypotheses have been advanced as to rodent origin and close relationships. The first allies rodents with primates (McKenna, 1961), and the second is the classical rodent-lagomorph relationship, at the basis of the concept Glires since the earliest classifications of mammals (for example Gregory, 1910). This latter hypothesis quickly became the prevailing one as more data and synapomorphic characters accumulated (Hartenberger, 1985; Luckett, 1977; Novacek, 1985).

2.1.2 Contribution of molecular data: Regression and progression of the debate

From a molecular point of view, the question of the systematic position of Rodentia among mammals is indisputably linked to the question of the order monophyly. As a matter of fact, in 1991, Graur et al. published a paper entitled "Is the guinea-pig a rodent?" that sounds as thunder in the peaceful life of rodentologists! This study was based on 15 protein sequences representing 1 998 aligned amino acids for four lineages: Primates, Artiodactyla , Rodentia (the guinea pig and a myomorph: rat or mouse or hamster) and one outgroup (marsupial or bird or toad). Maximum-parsimony analyses supported a tree in which rodents were not monophyletic because the guinea pig branched outside the clade ((Artiodactyla, Primates), Myomorpha). Later, this paper found an echo in "The guinea-pig is not a rodent" published by D'Erchia et al. (1996). This study was based on complete mitochondrial genome sequences of 16 taxa, representing five Primates, two Carnivora, two Cetartiodactyla, one Perissodactyla, one Insectivora, one Lagomorpha and three Rodentia (guinea pig, mouse and rat), with one Marsupialia as outgroup. Here again Rodentia did not appear monophyletic but contrary to the previous study, the clade *Mus-Rattus* appeared as the sister taxon to all other eutherian orders (except Insectivora). In this study the question of Glires was unsettled because the position of the rabbit was not robustly supported.

This challenge of rodent monophyly can retrospectively be considered as a textbook case, cumulating a number of molecular pitfalls that have since been well identified. First, the sampling question has immediately been underlined by some authors (Luckett & Hartenberger, 1993): the huge diversity of rodents (more than 2000 species) cannot be reduced to three taxa without consequences on phylogenetic inferences. This point was later confirmed by Lecointre et al. (1993) showing that reconstructed trees are highly sensitive to taxon sampling and that obtaining a reliable phylogeny necessitates to choose several taxa per presumed monophyletic lineage, as well as in the outgroups. Moreover, several authors (Cao et al., 1998; Philippe & Douzery, 1994) came to the conclusion that phylogenies based on four taxa (quartet analysis) can be highly misleading. The second problem concerned the methods used for phylogenetic inferences. Both studies refuting rodent monophyly were based on maximum parsimony analysis, a method already known to be subject to tree-reconstruction artefacts, such as "long branch attraction" (Felseinstein, 1978). Sullivan & Swofford (1997) reanalyzed the same datasets with more sophisticated probabilistic methods and showed the importance of using an adequate model of sequence evolution. According to their results, phylogenetic reconstructions can converge to a wrong tree, in particular if the model is oversimplified as in the studies of Graur et al. (1991) and D'Erchia et al. (1996). Among others, heterogeneity of substitutions between sites (modeled by a Gamma distribution) appeared as a particularly important factor to take into consideration for recovering a correct phylogeny.

After these two studies, more complete mitochondrial genomes were sequenced in rodents as well as in diverse mammalian orders (see references in Reyes et al., 2004). However, increasing the number of complete mitochondrial genomes or analyzing the data with probabilistic methods and adapted models did not change the first result: mitochondrial DNA proved to be unable to recover the monophyly of Rodentia (see Arnason et al., 2002). In all these studies, myomorphs (the rat and mouse lineage) appear as outside the rest of the rodents. Later the studies of Reyes et al. (2004) and Kjer et al. (2007) finally recovered a monophyletic Rodentia clade, probably because enough myomorph taxa were included to break the long branch leading to the Muridae (mouse and rat) family and also because effective methods such as Bayesian analysis and adapted models of sequence evolution were used. In the mean time, however, more and more studies based on nuclear genes invariably concluded that rodents are monophyletic (Adkins et al., 2003; Amrine-Madsen et al., 2003; DeBry & Sagel, 2001; Huchon et al., 2002). This result was usually achieved using extensive rodent sampling but also much less nucleotides than when whole mitochondrial genomes were considered. These contrasted results clearly showed that nuclear genes are much less affected by homoplasy than mitochondrial sequences, and thus appeared more appropriate to recover the phylogenetic signal for deep-level relationships (Springer et al., 2001). In fact, the debate was definitively closed in 2001 when two different studies (Murphy et al., 2001; Madsen et al., 2001) based on approximately 10 000 base pairs resolved most of the mammalian phylogenetic tree. Both papers came to the conclusion that Rodentia is a monophyletic group, to which Lagomorpha is the sister taxon, a result that finally reconciled morphologists and paleontologists with molecularists after 10 years of keen debate!

2.2 Rodent families and their phylogenetic relationships

2.2.1 The pre-molecular era: 100 years of work in diverse disciplines

Thirty-four living rodent families are currently recognized when including the Diatomyidae, a family recently reactivated (Huchon et al., 2007) to include *Laonastes aegnimamus* discovered in Laos in 1996 and described by Jenkins et al. (2005). Surprisingly, the number of rodent families has stayed relatively stable since the pioneer work of Brandt in 1855, i.e. approximately between 30 and 35 (Anderson & Jones, 1984; Hartenberger, 1985; Simpson, 1945; Wilson & Reeder, 1993, 2005). By contrast, interfamilial relationships were much more debated, which led to various proposals as to suprafamilial groupings. The earliest classifications of Brandt (1855) and Tullberg (1899) recognized two (Sciurognathi and Hystricognathi) and three (Hystricomorpha, Myomorpha and Sciuromorpha) major groups, respectively, but subsequent works identified more divisions (with the rank suborder, infraorder, or superfamily), the number of which ranged "anywhere from five to 16" (Carleton & Musser, 2005).

As early as in the Early Eocene, 11 families of rodent are already recognized (Hartenberger, 1998). Rodents most likely originated in Central Asia (Hartenberger, 1996) and within a few millions of years they diversified and dispersed on all continents to the exception of Antarctica and South America (Hartenberger, 1998). This radiation, that took place about 55 to 65 millions of years ago according to paleontological data or even earlier according to molecular data (between 70 and 80 Mya; Huchon et al., 2007; Montgelard et al., 2008), occurred so quickly that it was qualified as "explosive" (Hartenberger, 1996). Rapidly after their emergence, rodents also appeared ecologically diversified, and they currently occupy a tremendous variety of habitats in nearly all the ecosystems. This rapid geographical and ecological diversification is probably one of the reasons why numerous characters are homoplasic, precluding their use in phylogenetic studies. Before reaching this conclusion, a huge diversity of morphological and anatomical characters have been studied in the hope to discover that some of them escaped homoplasy: dental (Butler, 1985; Flynn et al., 1985; Marivaux et al., 2004) or cranial (Novacek, 1985) characters, cephalic arterial patterns (Bugge, 1985), middle-ear features (Lavocat & Parent, 1985), enamel (Martin, 1997) or placental characteristics (Luckett, 1985), paleontological data (Jaeger, 1988; Vianey-Liaud, 1985)... After 100 years of research, the consensual relationships were nevertheless very few. Well-supported evolutionary relationships concerned the close affinity between Geomyidae and Heteromyidae (Falbusch, 1985), Aplodontidae and Sciuridae (Lavocat & Parent, 1985; Vianey-Liaud, 1985; Wahlert, 1985), Dipodidae and Muroidea (a superfamily including six families; see paragraph 3) as well as the split of the Hystricognathi suborder in old world Phiomorpha (four families) and new-world Caviomorpha (13 families). Some putative affinities were also put forward, but less convincingly, such a sister group relationship between Gliridae and Sciuridae + Aplodontidae (Bugge, 1985; Lavocat & Parent, 1985), Ctenodactylidae and Hystricognathi (Jaeger, 1988; Luckett, 1985) or Anomaluridae and Pedetidae (Bugge, 1985; Lavocat & Parent, 1985). Other proposed relationships were more speculative (see Luckett & Hartenberger, 1985) and a number of families stayed as *incertae sedis* because of inconclusive results. Finally, no strong hypothesis has ever been put forward concerning suprasubordinal relationships or the position of the root of the rodent tree based on traditional morpho-anatomical characters.

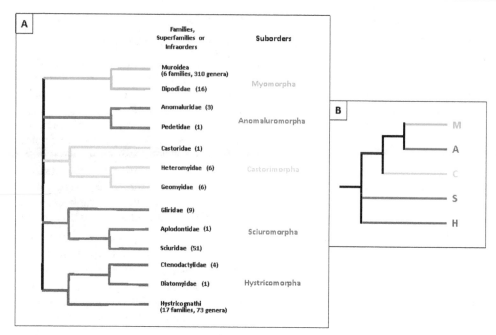

Fig. 2. A: The 34 rodent families and their relationships in five suborders (Carleton & Musser, 2005); B: Likely phylogenetic arrangement of the five suborders in three major lineages (Blanga-Kanfi et al., 2009 ; Montgelard et al., 2008).

2.2.2 Contribution of molecular data: Reaching a consensus in five clades

In this context of uncertainty about suprafamilial relationships, the contribution of molecular data was essential, not only for selecting between the diverse prevailing hypotheses but also because new relationships emerged that had never been advanced by previous morphological or paleontological analyses. The first studies were based on mitochondrial genes (cytochrome b and 12S rRNA; Nedbal et al., 1996; Montgelard et al., 2002) but several nuclear genes have then been sequenced intensively: the von Willebrand Factor (vWF; Huchon et al., 1999), the interphotoreceptor retinoid-binding protein (IRBP; Huchon et al., 2002; DeBry, 2003), the alpha 2B adrenergic receptor (A2AB; Huchon et al., 2002), the growth hormone receptor (GHR; Adkins et al., 2003), a breast and ovarian cancer susceptibility gene (BRCA1; Adkins et al., 2003), or the apolipoprotein B (APOB; Amrine-Madsen et al., 2003), among others. Contrary to what happened at the beginning of the molecular era (see above), all these molecular studies included a substantial number of rodent taxa and were mostly congruent (at least for the groupings strongly supported). In particular, all well-supported associations previously mentioned (see paragraph 2.2.1 and figure 2) were confirmed, allowing to invalidate some alternative morpho-paleontological propositions such as for example Ctenodactylidae as the first emerging group among rodents (Hartenberger, 1985) or a group including Muridae, Dipodidae, Heteromyidae and Geomyidae (Myomorpha *sensu* Carleton, 1984). Even if the phylogenetic signal appears less altered in nuclear than in mitochondrial genes (see Springer et al., 2001), these relationships were often evidenced by the combination of several genes.

Finally, most recent molecular studies on rodent phylogeny (Blanga-Kanfi et al., 2009; Huchon et al., 2007; Montgelard et al., 2008) agree on the identification of five major clades among Rodentia (Figure 2A): Myomorpha (Muroidea and Dipodidae), Anomaluromorpha (Anomaluridae and Pedetidae), Castorimorpha (Geomyidae, Heteromyidae, Castoridae), Sciuromorpha (Aplodontidae, Sciuridae and Gliridae) and Hystricomorpha (Hystricognathi, Ctenodactylidae and Diatomyidae). These groupings, which are given suborder ranks in Carleton & Musser (2005) are very dissimilar in terms of taxonomic diversity: from only two families and four genera for Anomaluromorpha to seven families and 326 genera for Myomorpha which represents nearly 70% of rodent diversity.

2.3 Relationships between the five major clades of Rodentia

2.3.1 Influence of factors of inconsistency on rodent suprafamilial relationships

As stated in the introduction, molecular phylogeny can be subjected to different biases and lead to erroneous trees, such as described about the monophyly of rodents (see paragraph 2.1.2). A number of factors have been identified as major sources of inconsistency (Philippe et al., 2005): variation in nucleotide composition, across-site heterogeneity, shifts in the evolutionary rate among lineages, or heterotachy (rate variation across a site through time). Two studies (Blanga-Kanfi et al., 2009; Montgelard et al., 2008) included representatives of all rodent families and analyzed the influence of different factors of inconsistency on the relationships between the five major clades of rodents. The paper of Blanga-Kanfi et al. (2009) is based on six nuclear protein-coding genes (6 255 positions) whereas the paper of Montgelard et al. (2008) used two mitochondrial genes (2 126 sites), two nuclear exons (2 571 sites) and four nuclear introns (2 897 sites). With the aim to increase the ratio between phylogenetic and nonphylogenetic signal, both studies analyzed different methods for thwarting factors of inconsistencies, among which only the effect of removing fast-evolving positions will be reported here. In the two studies, the five major rodent clades are highly supported as well as a super-clade including Myomorpha + Anomaluromorpha + Castorimorpha. This grouping has never been suggested from morpho-anatomical or paleontological data and is referred as "the Mouse-related clade" (Huchon et al., 2002) pending a true binomial denomination. Inside this clade, a sister group relationship between Anomaluromorpha and Myomorpha is privileged over the other two possibilities. This clade is highly supported by the intron data in the study of Montgelard et al. (2008) whereas other datasets (including the six nuclear genes of Blanga-Kanfi et al., 2009) only moderately supported this relationship, whatever the analyses considered. From there, rodent phylogeny at the highest taxonomic level can thus be reduced to three major lineages: Sciuromorpha, Hystricomorpha and the Mouse-related clade (Figure 2B).

The next step was to try resolve the order of divergence between these three lineages, that is to identify the root of Rodentia. The study by Blanga-Kanfi et al. (2009) does not support any of the three possibilities (one of the three lineages at the base of rodents) when all the nucleotide dataset is considered and whatever the analyses performed (different types of models). Conversely, when fast-evolving positions are removed from their complete dataset (1114 nucleotides deleted on 7594 deleted), Montgelard et al. (2008) obtained a strong support for the clade Hystricomorpha + Sciuromorpha, to which the Mouse-related clade thus showed a basal position. By contrast, removal of the fastest-evolving sites (258 positions on 6255) moderately improved the support for a basal position of Sciurmorpha in the paper of Blanga-Kanfi et al. (2009) but this position is reinforced when the protein sequences are analyzed with sophisticated evolutionary models.

We can finally report the study of Churakov et al. (2010) who addressed the question of the root of the rodent tree on the basis of the insertion of SINEs (short interspersed repetitive elements). These are mobile elements whose insertion is considered as a unique and irreversible event at the genome scale. Thus they are regarded as virtually free of homoplasy, making SINEs efficient tools for reconstructing phylogenetic relationships (Serdobova & Kramerov, 1998). Concerning Rodentia, Churakov et al. (2010) found eight retroposon insertions and two indels as characterizing the clade Hystricomorpha+Mouse-related, thus making Sciuromorpha the first offshoot among rodents, in accordance with Blanga-Kanfi et al. 's (2009) results. However, the hypothesis of a basal position of the Mouse-related clade is not totally refuted as they also identified two SINES and one indel shared by Hystricomorpha and Sciuromorpha. The authors put forward two hypotheses to explain these data: an incomplete lineage sorting or introgressive hybridization occurring in the early stages of these two main rodent lineages. Another hypothesis, not invoked by Churakov et al. (2010) is that SINES would be homoplasic as already described in the literature (Nishihara et al., 2009).

2.3.2 Future prospects for molecular, morphological and paleontological data

Concerning molecular data, we could be tented to conclude that relationships between the three main rodent lineages would represent a real hard polytomy (that is they diverged simultaneously) because as much as 6000 characters cannot definitely solved the relationships between them. However, it could also be that the nucleotide dataset considered is not large enough. Indeed, 16 kb of DNA sequences have been necessary to solve the phylogenetic relationships between the different families of Madagascar's lemurs (Horwath et al., 2008) and 24 000 nucleotides succeeded to fix ratite relationships (Harshman et al., 2008)!

As to anatomical, morphological or paleontological data, their future input to phylogenetical and systematic studies likely lies in a reanalysis of characters in the light of molecular phylogenies with the major challenge to understand morphological evolution. Moreover, it should not be forgotten that very important technical progresses have been performed (such as geometric morphometric methods) allowing the quantification of shape modifications (see for example Hautier et al., 2011). In the current frame of knowledge of higher-level rodent relationships, studies can now be conducted with the scope to understand if characters common to different lineages do represent real homologies (common ascendance) or not. In this context, fossils could constitute temporal landmarks in order to trace morphological modifications. Finally, the advancement of new disciplines such as evolutionary developmental biology ("evo-devo" in short) will probably link molecular and morphological disciplines, allowing to understand the respective roles of ascendance (phylogeny) and ecology (selection) during evolution of morphological characters (see for example Renvoise et al., 2009).

3. Intermediate taxonomic levels (families – subfamilies – tribes) of the Muroidea superfamily

Within each of the five rodent suborders, a wealth of molecular studies based on DNA sequences has been devoted to determine the systematic arrangement around the familial level. As an example, we will examine the systematic relationships among the superfamily Muroidea, which includes six families (Calomyscidae, Platacanthomyidae, Spalacidae, Nesomyidae, Cricetidae and Muridae), and represents by far the most speciose group of rodents (Figure 3). Muroidea has been the focus of a number of molecular studies based on various mitochondrial and nuclear genes (Jansa et al., 2009; Jansa & Weksler 2004; Michaux & Catzeflis, 2000; Michaux et al., 2001; Steppan et al., 2004). Figure 3 schematically illustrates the most probable relationships between all the families and subfamilies currently recognized in this superfamily. This arrangement served as a basis for the systematic classification retained by Musser & Carleton (2005), and resumed in Figure 3 (see * in the legend for departures to this arrangement). We will here review, for each family, their content and organization on the basis of the different molecular analyses that have been performed. It can also be mentioned that these studies had for consequence the reanalysis of dental character evolution in the light of molecular phylogenies. For example, the study by Lazzari et al. (2008) highlighted the weight of functional constraints and revealed numerous dental homoplasies among the different morphological grades observed in the course of muroid evolution.

3.1 Calomyscidae and Platacanthomyidae

As compared to other muroid families (see Figure 3), Calomyscidae (1 genus, 8 species) and Platacanthomyidae (2 genera, 2 species) are small families and only a few studies integrated them in large-scale analyses to precise their phylogenetic position within the Muroidea. In the case of the Asian family Platacanthomyidae, a basal position has been evidenced for *Typhlomys* among muroid rodents, a result that led Jansa et al. (2009) to propose a Eurasian origin for the Muroidea. Calomyscidae appeared as an isolated lineage in all molecular studies (Jansa et al., 2009; Jansa & Weksler 2004; Michaux et al., 2001; Steppan et al., 2004). Molecular data clearly grouped Calomyscidae with Dendromuridae, Muridae and Cricetidae, but no study has yet resolved the branching pattern within this group

3.2 Spalacidae

In all phylogenetic reconstructions based on molecular data, fossorial Spalacidae appeared as an early differentiated lineage in the Muroidea, of Eurasian origin (Jansa et al., 2009; Jansa & Weksler, 2004; Robinson et al., 1997; Steppan et al., 2004). The content and internal relationships of Spalacidae have long been debated (see Gogolevskaya et al., 2010 for details), until Jansa & Weksler (2004) and Norris et al. (2004) recognized Myospalacinae, Rhizomyinae, and Spalacinae as distinct, but closely related subfamilies of Spalacidae within the Muroidea based on nuclear as well as mitochondrial sequence comparisons. This result was later confirmed by Gogolevskaya et al. (2010) who showed that representatives of these subfamilies shared the same variants of small genetic sequences, namely the B1 small interspersed elements (SINEs) and the $4.5S_I$ small nuclear RNA. This study confirmed the role of SINES in rodent phylogeny, as suggested by Serdobova & Kramerov (1998), already using Spalacidae. The distinctiveness of Tachyoryctinae (African mole rats), especially with respect to Rhizomyinae (Asian bamboo rats), is currently mainly supported by morpho-anatomical arguments (Musser & Carleton, 2005). On the other hand, molecular data gathered to date rather showed the close relatedness of *Rhizomys* and *Tachyoryctes*, that both Jansa & Weksler (2004) and Steppan et al. (2004) considered as belonging to the Rhizomyinae, an hypothesis that would mean the disappearance of the Tachyorictinae subfamily as a taxonomic rank. Finally, Norris et al. (2004) and Jansa et al. (2009) evidenced

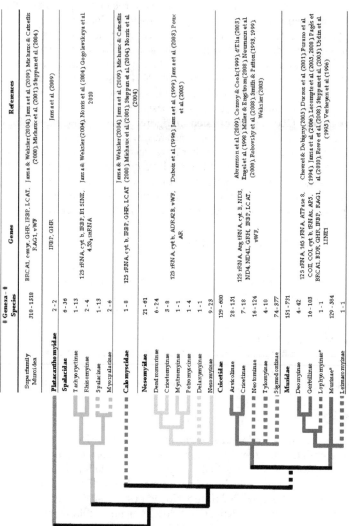

Fig. 3. Strongly supported (in bold) and hypothetic/conflictual (dotted lines) phylogenetic relationships between families / subfamilies of Muroid rodents as established following various studies using DNA sequences. * Otomyini included in Murinae, and Lophyomyinae included in Muridae, as in Jansa & Weksler (2004) / Jansa et al. (2009). 4.5SI snRNA = 4.5SI small nuclear RNA; ADRA2B = alpha 2B adrenergic receptor; AP5 = Acid Phosphatase type V; AR = androgen receptor; Arg tRNA = Arginine tRNA; B1 SINE = B1 short interspersed element; BDR = Borrelia direct repeat; BRCA1 = Breast Cancer Gene 1; c-myc = proto-oncongene; COI and COII: Cytochrome Oxydase subunits I and II, cyt b = cytochrome b; GRH = Growth Hormone Receptor; IRBP = Interphotoreceptor Retinoid Binding Protein; LCAT = Lecithin Cholesterol Acyl Transferase; LINE1 = Long Interspersed repeated DNA; ND3, ND4, ND4L = subunits of nicotinamide dinucleotide dehydrogenase; RAG1 = Recombination Activating Gene 1; vWF = von Willebrand Factor

a sister group relationships between Myospalacinae and Spalacinae, an affinity that is however not strongly supported.

3.3 Nesomyidae

The current contours of the family Nesomyidae owe much to molecular data. Six subfamilies (Cricetomyinae, Delanymyinae, Dendromurinae, Mystromyinae, Nesomyinae, and Petromyscinae) are currently recognized, a result that echoes some of the hypotheses brought forward by Lavocat (1973, 1978) based on palaeontological evidence. Some Malagasy endemic Nesomyinae were first found related to *Cricetomys*, an African representative of the Cricetomyinae (Dubois et al., 1996) based on 12S rRNA mitochondrial gene. The results obtained by Jansa et al. (1999) were more ambiguous as to the relationships between Nesomyinae and other muroid subfamilies, probably because they were based on cytochrome *b*, a too rapidly evolving gene for addressing adequately issues related to such ancient events. Indeed, subsequent studies by Jansa & Weksler (2004), Michaux & Catzeflis (2000), Michaux et al. (2001) and Steppan et al. (2004), all based on nuclear genes, clearly showed that the Nesomyidae as defined here above represent a well supported natural group. Within it, the Malagasy endemic Nesomyinae would be the sister group to a clade including two pairs of taxonomically equivalent groups, namely the Cricetomyinae and Dendromurinae on the one hand, and the Mystromyinae and Petromyscinae on the other hand (see Jansa & Weksler, 2004; Jansa et al., 2009). The Delanymyinae, with only one secretive species, has never been involved in phylogenetic analyses based on molecular data, and its inclusion within the Nesomyidae mainly rests on its previous association with either Dendromurinae or Petromyscinae based on morpho-anatomical characters (Musser & Carleton, 2005).

Studies focusing on Nesomyidae subfamilies mainly concerned Nesomyinae and Dendromurinae. In the former, Jansa et al. (2009) and Jansa & Carleton (2003) privileged the hypotheses of a unique event of colonization of Madagascar by this endemic group, an hypothesis congruent with the monophyly of Nesomyinae regularly evidenced using nuclear gene sequences (Steppan et al, 2004; Poux et al, 2005). Dendromurinae was the topic of various molecular studies which, together with morphological re-analyses, finally led to precise its generic content (see Musser & Carleton, 2005). In particular, Verheyen et al. (1996) using mitochondrial cytochrome *b* sequences, then Michaux & Catzeflis (2000) using the nuclear gene LCAT, definitely proved that *Deomys* should not be considered as a member of the Dendromurinae, confirming earlier results obtained by Denys et al. (1995) based on morphological and DNA-DNA hybridization data. As currently understood, the Dendromurinae subfamily now comprises the species-rich genera *Dendromus* and *Steatomys* as well as the monospecific genera *Dendroprionomys, Malacothrix, Megadendromus,* and *Prionomys.*

3.4 Cricetidae

The family Cricetidae has long been considered as a subgroup of an extended Muridae family, and is still a matter of controversy as far as its limits are concerned (Musser & Carleton, 2005). From a molecular perspective, earlier studies were also quite hesitant in their conclusions, as various lineages of current Cricetidae and Muridae appeared to have emerged nearly simultaneously, giving polytomies that were interpreted as resulting from adaptive radiations (see Conroy & Cook, 1999; Michaux & Catzeflis, 2000). Jansa & Weksler

(2004) and Steppan et al. (2004), based on nuclear gene sequences and a relatively comprehensive taxonomic sampling, evidenced a monophyletic group corresponding to Cricetidae in their "modern" meaning, which includes the subfamilies Arvicolinae, Cricetinae, Neotominae, Sigmodontinae, and Tylomyinae. If each subfamily appeared well supported by molecular data, affinities between them are uncertain: A relatively poorly supported clade joining the Holarctic Arvicolinae and the Palaearctic Cricetinae was found by Steppan et al. (2004) based on a combination of nuclear gene sequences. Alternatively, Jansa et al. (2009) suggested a sister relationship between Sigmodontinae and Neotominae, and a basal position of Cricetinae among all other Cricetidae but here again with relatively low support.

The content and organization of Arvicolinae has since then been the subject of many molecular studies, including those by Abramson et al. (2009) and Robovsky et al. (2008), who tried to arrange the numerous tribes constituting this species-rich subfamily. Interestingly, the distribution of satellite DNA also proved to be informative at the infra-subfamilial level, bringing support to the tribe Arvicolini (Acosta et al., 2010). The Cricetinae have been studied via mitochondrial and nuclear sequences by Neumann et al. (2006) who evidenced a well-supported phylogenetic structure in three main lineages that would have diverged during the late Miocene (7-12 Myr ago). Sigmodontinae, the New World Cricetidae, were also shown to comprise three monophyletic clades by Jansa & Weksler (2004) and Steppan et al. (2004), a finding that was already apparent in an earlier study by Engel et al. (1998) based on various mitochondrial genes. Among them, Sigmodontinae "*sensu stricto*" represent the South American offshoot of the family, whose colonization between 5 and 9 Myr ago of the subcontinent from North America is a remarkable example of the filling of an empty niche following a fortuitous invasion (Engel et al., 1998). The internal organization of this highly diverse subfamily (> 300 species) has been studied using mitochondrial and nuclear gene sequences by various authors, including Smith & Patton (1993, 1999), d'Elia et al. (2003), and Weksler (2003). The Neotominae mostly represent the North American branch of the Cricetidae, among which the so-called Peromyscine rodents is by far the most speciose group. Through a combination of mitochondrial and nuclear sequences, Miller & Engstrom (2008) ended up to a well-resolved phylogeny confirming the Reithrodontomyini as the major tribe of this subfamily, with Baiomyini, Ochrotomyini and Neotomini as successive sister taxa. Previously, Bradley et al. (2004) based on cytochrome *b* gene data, quite convincingly showed that the Tylomyinae (then treated as the tribe Tylomyini) appeared as basal to the other Neotominae considered.

3.5 Muridae

A monophyletic group equivalent to the Cricetidae (see above) emerged in molecular phylogenetic analyses conducted at the Muroid level (Jansa et al., 2009; Jansa & Weksler, 2004; Michaux et al., 2001), that was proposed as the Muridae family by Steppan et al. (2004). The combination of molecular results and other sets of characteristics (Musser & Carleton, 2005) led to organize this speciose assemblage into five main subfamilies (Leimacomyinae, Deomyinae, Gerbillinae, Lophyomyinae, Murinae). Leimacomyinae only comprise the monospecific genus *Leimacomys*, known by two specimens caught in 1890. Its inclusion in the Muridae still awaits confirmation, especially from a molecular perspective. Deomyinae (corresponding to the former Acomyinae, see Musser & Carleton, 2005 for details) emerged following various molecular analyses (including DNA-DNA hybridization experiments; Chevret et al., 1993) which confirmed that i) *Acomys* should not be considered as a Murinae (see Dubois et al., 1999 for instance), and ii) *Deomys* was closely related to *Acomys*, but also to *Lophuromys* and *Uranomys* (Michaux et al., 2001; Steppan et al., 2004, 2005; Verheyen et al., 1996). These four genera comprise a well-supported clade which should be named Deomyinae, with the Gerbillinae as sister group. The latter subfamily represents a rather homogeneous group from a morpho-anatomical point of view, whose content was not adequately surveyed using molecular markers before Chevret & Dobigny (2005). Based on mitochondrial gene sequences, these authors specified the generic content of the group, identified three main clades that may correspond to distinct tribes, and suggested an African origin for the Gerbillinae, with subsequent migration events towards Asia. The monospecific Lophyomyinae, considered as belonging to the Cricetidae based on skull and dental characteristics (Musser & Carleton, 2005), is here placed within the Muridae, following Jansa & Weksler (2004) who found it to be the sister group to the Deomyinae + Gerbillinae clade, although this relationship would need to be strengthened.

The Murinae represents a huge assemblage of species (Figure 3). For this reason, it has seldom been considered exhaustively in molecular works. However numerous studies focused either on taxonomical or geographical subgroups of murine rodents, progressively leading to significant advances in the systematics of this subfamily. Nuclear and mitochondrial sequence data were predominantly used, but interspersed repeated DNA (LINE-1 or Lx family) also proved to be useful as heritable characters in defining the murine lineage or some of its sub-parts (Furano et al., 1994; Usdin et al., 1995). Steppan et al. (2005), based on separate and combined analyses of mitochondrial and nuclear genes sequenced in most major murine groups, showed the basal split of a clade of Philippine old endemics (corresponding to Phloeomyini *sensu* Lecompte et al., 2008), the remaining taxa being organized within at least 7 geographically structured lineages. Among them, a South-east Asian "*Rattus*" clade (Rattini tribe *sensu* Lecompte et al., 2008) has recently been analyzed by Pagès et al. (2010) in a thorough phylogeny-based taxonomic revision. Another of these lineages, known as the Sahul (Australia + New Guinea) old endemic rodents, appeared as the sister group to another lineage of Philippine old endemics (Jansa et al., 2006; Rowe et al., 2008). Both clades probably derived from a single colonization event of New Guinea from the West, during the Late Miocene – Early Pliocene period. Lecompte et al. (2008) proposed the tribal name Hydromyini for this entire assemblage including the Australo-Papuan and Philippine murine radiations. Ducroz et al. (2001), using complete cytochrome b, and partial 12S and 16S ribosomal RNA mitochondrial sequences, evidenced a major clade of mostly African genera to which they proposed the tribal name Arvicanthini. Another important lineage of African rodents was defined as "the *Praomys* group" by Lecompte et al. (2005), before being formally named Praomyini (Lecompte et al., 2008). A number of other, either Eurasian (Apodemini, Millardini, Murini) or African (Otomyini, Malacomyini), tribes were also advanced by Lecompte et al. (2008). Overall, these authors proposed that the Murinae diversity be organized in at least 10 tribes, and, given the pattern of relationships observed between them, that multiple exchanges occurred between Eurasia and Africa in this subfamily. The first colonization event of Africa would have taken place around 11 Myr ago, followed by a major period of diversification between 7-9 Myr ago.

As apparent following this rapid survey, significant progress has been made within the last 15-20 years regarding the content of, and pattern of relationships between, the main muroid rodent lineages. This result has been achieved through the use of adequate genetic markers coupled to large taxonomic sampling representative of the diversity of the groups under scrutiny. Beside their purely taxonomic outputs, these studies have also yielded a wealth of knowledge on the biogeographical history of muroid rodents throughout the World. From a temporal point of view, the basal radiation of Muroidea would have taken place around 25 Myr ago, then major lineages (families and subfamilies) progressively differentiated during the following 10-12 Myr (Steppan et al., 2004). The generic diversification occurred at various periods, but was clearly enhanced (adaptive radiations) when new spaces were colonized, such as South America and Africa. This mainly Miocene story has taken place over nearly all the continents, with migration/colonization events that proved to be crucial in the evolution of the group towards its current diversity.

4. The species level in rodent taxonomy

Until recently, systematic and taxonomic proposals based on sequence data at the species level often arose as a by-product in studies focusing either on phylogenetic relationships or evolutionary processes of a given supraspecific group. However, molecular data now tend to be integrated in multidisciplinary studies explicitly devoted to species characterization / description. This trend especially develops in groups where cryptic and sibling species are numerous, and where the use of genetic markers rapidly proved to be of paramount importance. This was accompanied by a conceptual evolution in the field of species concepts that now take into account molecular and other genetic characters in species delimitation. From a practical point of view, the use of sequence data has also prompted a debate on how genetic distances and genetic characters should be considered in species delimitation and description. These reflections have logically resulted in the recent proposals centred on the idea of DNA taxonomy, among which DNA barcoding has gained most of the attention.

4.1 Contribution of sequence data to cryptic and sibling species identification

As defined by Knowlton (1986), sibling species represent a particular class of cryptic species, i.e. those that are phylogenetically closely related (often sister species). These morphologically very similar species have for long been identified as a major obstacle to the application of the morphological species concept (see Mayr, 1948). The increasing recognition of cryptic taxa, undoubtedly linked to the generalization of genetic tools (including DNA sequencing), poses various questions concerning their distribution across taxa and biomes, their conditions of emergence, and their impact on biodiversity estimates (Bickford et al., 2007).

Patterson (2000) reviewed recent discoveries in Neotropical mammals, and highlighted the fact that among the "newly recognized species", the re-evaluation of already collected and studied materials outnumbered real *de novo* descriptions. According to Patterson (2001: 195), this trend to resurrect synonyms is jointly attributable to "continued morphological study, higher resolution genetic analyses and a shift toward a phylogenetic species concept (and away from polytypic species)". Rodents represent sixty percent of these new species, i.e. a much greater proportion than their actual share in the Mammalia class (ca. 40% of genera or species). Given the current activity of description / recognition of rodent species in Asia, one may hypothesize that the same trends would be found on this continent. In Africa, rodents and primates are the two mammalian groups in which the largest numbers of new species have been described in the decade 1989-2008 (ca 33% each; Hoffmann et al. 2009). In both groups, numerous of these novelties have been characterized thanks to genetic methods, including cytogenetics (Taylor, 2000; Granjon & Dobigny, 2003) and sequence-based analyses (see hereafter).

In African rodents, beside re-evaluation of some groups using traditional methods, molecular phylogenetic studies have opened the way to many species resurrections, particularly in speciose groups such as the "*Arvicanthis* division" (Ducroz et al., 2001), the *Praomys* group (Lecompte et al., 2002, 2005) or the genus *Lophuromys* (Verheyen et al., 2007). Phylogeographical studies are also the occasion for taxonomic revisions, including the finding of cryptic and sibling species, as acknowledged by Avise (2000: 204). The recent work by Bryja et al. (2010) on the *Praomys daltoni* complex both confirmed the synonymy of *P. daltoni* and *P. derooi* and strongly suggested the existence of a new, sibling species for this West African murine rodent. It has to be underlined that the "lumping" of taxa previously thought to represent distinct species could also result from phylogeographical surveys, as in Nicolas et al. (2009) where the wide-ranging deomyine *Acomys airensis* was shown to represent a junior synonym of the localized (but previously described) *A. chudeaui*.

4.2 Genetic clusters and species concepts

Species concepts have multiplied along the history of evolutionary sciences, from the unique typological and mainly morpho-anatomically based species concept to a variety of concepts reflecting both the discipline of predilection and the school of thoughts of their promoters (see the useful reviews by Harrison [1998] and De Queiroz [2007, 2011]). Among them, those referenced as the "genealogical" and the "diagnosable" species concepts are both considered as phylogenetic by De Queiroz (2007). They are also strongly linked to the development of molecular studies because exclusive coalescence of alleles and diagnosability (via qualitative, fixed difference), are considered as the main properties of species. One of the main practical problems produced by this proliferation of species concepts is the variability in the number of species inferred by any of them. As exemplified by De Queiroz (2011), the criterion of fixed character state differences (the basis of phylogenetic species concepts) commonly leads to the recognition of more species taxa than the criterion of intrinsic reproductive barriers (the basis of the biological species concept). A shift toward a phylogenetic species concept was one of the reasons invoked by Patterson (2001) to explain the current trend in Neotropical species number increase (see above). The "genetic species concept" of Dobzhansky (1950), revisited by Bradley & Baker (2001) and Baker & Bradley (2006) even goes further, considering as genetic species all genetically defined phylogroups, especially when based on DNA sequence analyses. This concept, which explicitly focuses on genetic isolation rather than on reproductive isolation, would mainly concern morphologically non-differentiated species. Strictly applied, it might lead to an increase of >2,000 species in mammals only (Baker & Bradley, 2006).

Hopefully, after a period of more or less anarchic burgeoning of such "specialized" species concepts, the current trend is now to try reconciling them into a unified species concept where species are considered as "evolving metapopulation lineage" that can be diagnosed via one or (better) several properties. These properties could be drawn from any discipline, provided it will furnish convincing lines of evidence of species delimitation (De Queiroz, 2007). In this frame, molecular data, when correctly interpreted, can of course represent strong arguments in favour (or not) of the recognition of an independent lineage as a true species. Interestingly, this new approach emerges at the same time as integrative systematics or taxonomy (Dayrat, 2005; Lecompte et al., 2003) that recommend the integrated use of data from various fields in order to adequately and critically assess hypotheses on species delimitation and characterization. However, when diagnostic characters have to be identified, which is one of the main tasks of taxonomists, sequence data may present some inherent difficulties. This inconvenience should be overcome, or at least critically taken into account when the corresponding results have to be interpreted in a systematic context (see hereafter).

4.3 Genetic distances and characters: Their practical use in systematics and taxonomy

The use of genetic data to delimitate species has for long been a matter of debate. In the 70's and 80's when protein eletrophoresis was the main tool used to compare genetic diversity within and between species, genetic distances were generally computed, and interpreted as measuring the level of differentiation of populations / species. Correlations between such distances and the taxonomic ranks of the groups under study were examined. Important overlaps were then observed between genetic distance ranges and taxonomic levels (from subspecies to well-differentiated species via sibling species), that preclude definitive taxonomic conclusions based on such correlations (Zimmerman et al., 1978; Avise & Aquadro, 1982; Graf, 1982). Similarly, when sequence data have started accumulating, various authors have compiled sequence divergence values (especially K2P [Kimura two parameter] distance from cytochrome b gene data) *versus* taxonomic ranks, showing the same pattern of overlap (see Bradley & Baker, 2001 for Mammals; Johns & Avise, 1998 for Vertebrates). As acknowledged by various authors, and recently summarized by Coleman (2009: 197): "Nucleotide change does occur, genera differ more than do their component pairs of species, but the nucleotide change is continuous, with no gap, no point of reference correlatable with some facet of speciation." Under the biological species concept, where the achievement of reproductive isolation is the major criterion to define species, the use of such distance data appears widely equivocal (Ferguson, 2002): recently diverged species (showing reproductive isolation) can exhibit a smaller genetic divergence than conspecific populations genetically differentiated (e.g. because of geographic distance), but still reproductively compatible. Conversely, a small genetic distance between two individual sets of any given taxon, even if they show statistically well-supported reciprocal monophyly, may cast doubt on the distinct specific status of these groups in absence of other convincing diagnostic criteria (see Ferguson, 2002 for some examples). These criteria should better be structural (morpho-anatomical, karyological...), whereas size or shape variation may represent inadequate characters, being often subject to local selection linked with geographical, environmental or biotic factors. Thus, while recommending the general procedure followed by Gündüz et al. (2007) who considered molecular subdivision as an indicator of reproductive isolation for recognizing species boundaries, then used geometric morphometrics and external morphology to assess the amount of phenotypic partitioning among the species identified, we nevertheless insist on the importance of putting forward unambiguous, selection-free, diagnostic characters to characterize new species.

The identification of molecular apomorphies (or "fixed genetic characteristics", Ferguson, 2002) could represent an alternative to the sole use of genetic distance to characterize a new species. This was achieved, among others, in the recent rodent species descriptions by Pardiñas et al. (2005), Goodman et al. (2009), and Jayat et al. (2010). However, as underlined by the latter authors themselves, these character states should be taken with caution when sequences of some related species are not included, and/or when only a few haplotypes of the new species are considered. Another perspective, proposed by Coleman (2009) in a "biological species concept" frame, would be to use genes either involved in the reproductive barrier (sexual behaviour, gamete approach / fusion…), or for which sequence variation would accompany the separation of clades into sexually isolated subclades. The Internal Transcribed Spacer 2 region (ITS2) of the nuclear ribosomal cistrons, one of the most studied nuclear species-level molecular marker (Hajibabaei et al., 2007), seems a promising candidate in this respect. This gene displays both sequence and secondary structure variations that are highly correlated with taxonomic classification. Thus, it might represent a powerful tool because not only sequences can be compared but also some aspects of the secondary structure formed by the initial RNA transcript (Coleman, 2009).

4.4 DNA barcoding and DNA taxonomy

The concomitant loss of taxonomic expertise, the need for large-scaled biodiversity evaluation, and the development of both molecular techniques and computing power have prompted the emergence of DNA taxonomy, where nucleotide sequences are, as a first approximation, taken as diagnostic characters of the species under study (Blaxter, 2003, 2004). Within this frame, the objective of DNA barcoding is to provide a simple diagnostic tool, based on the correspondence between DNA sequences (generally of the cytochrome oxydase I mitochondrial gene) and species as defined via traditional systematics (Hebert et al., 2003). The progress of DNA barcoding, and its contribution to the description of species, especially of cryptic and sibling species have been acknowledged by Frézal & Leblois (2008). This study however, identified several crucial pitfalls, such as those linked to the representativity of species diversity on the one hand, and those associated with the peculiarities of mitochondrial DNA (maternal inheritance, risk of nuclear copies, variations in rate of evolution across taxa) on the other hand. The necessity for adequate sampling (of taxa within a given group and of individuals within a species), and for sequencing of nuclear markers, appear as useful recommendations to overcome these potential biases (Frézal & Leblois 2008). Enlarging sample sizes may also help to distinguish between intraspecific variations and species-level dichotomies. Pons et al. (2006) addressed this point in their "general mixed Yule coalescent" (GMYC) model, which combines models of stochastic lineage growth with coalescence theory to develop a new likelihood method that determines the point of transition between species-level (interspecific long branches) and population-level (short burgeoning branches) evolutionary processes. Pagès et al. (2010) recently adopted this approach, which does not require defining entities *a priori*, for the murine rodent tribe Rattini, a group in which species identification is difficult to assess through morphological determination. Pagès et al. (2010) thus recovered 24 putative species, to which they could *a posteriori* attribute an unambiguous species name to 18 of them. The remaining six groups either corresponded to small samples for which insufficient data was available to choose a valid name, or to potentially new species. Hence, although being in essence derived from purely molecular procedures, this phylogeny-based method for delimiting species also includes an integrative taxonomy approach when it comes to taxa characterization and naming.

4.5 Sequence data and recent species description

To check for the importance of molecular data in recent descriptions of new rodent species, two distinct datasets were analyzed. First, the list of muroid rodent species described between 1990 and 2003 was drawn up from Musser & Carleton (2005). Second, all scientific papers describing new rodent species published between 2004 and 2011 were tentatively gathered. In both cases, the relative importance of DNA sequence data in the characterization of new species was evaluated. In the description of new species, morphological information still has a major place, as a detailed description of type series (and especially of the holotype) is required as well as a comparison of the new species structural characteristics relative to related species. However, information from additional fields are now regularly added, that can be of prime importance in the new species diagnosis.

One hundred and seven species of muroid rodents were described between 1990 and 2003, according to Musser & Carleton (2005). Most of them (i.e. > 94%) belong to the families Cricetidae (N = 52) and Muridae (N = 49), in agreement with their importance within the superfamily Muroidea, where they represent nearly 94% of the total number of species (see Figure 3). Musser & Carleton (2005) quoted some of these new species as debatable because either based on a very small number of specimens (sometimes only one), or in need of further and more detailed comparison with already existing species. Morphological and, most of the time, morphometrical data constitute the basis of these descriptions, although in a few cases this information appears hardly useful, or even useless, for species diagnosis. This is the case for instance for *Taterillus tranieri*, an example of sibling species whose main characteristic consists in its karyotype (Dobigny et al., 2003). It is noteworthy that chromosomal features represented the most frequently included character (in addition to morphological ones) in species descriptions, being present in at least 31 of the cases (i.e. nearly 30%). This highlights the importance of karyotypic data as species-specific diagnostic characters in rodents, a group where chromosomal variability has already been underlined in several occasions (see Granjon & Dobigny, 2003 and Taylor et al., 2000, for recent examples in African rodents). By contrast, DNA sequence data were only provided as a support to the new species description in 10 cases (i.e. 9.3%). Musser & Carleton (2005) mentioned that in at least 8 other instances, molecular data came soon after species descriptions, generally confirming the validity of the concerned species. Interestingly, protein electrophoresis data were still used in a small number of these descriptions (less than 10).

In the period 2004-2011, we identified 55 new rodent species descriptions in the literature (Table 1). Although all regions of the world are concerned, the tropical belt concentrates most of these biological novelties. Among them, 20 concerned the family Cricetidae and 26 the family Muridae. Sequence data proved to be directly involved in 30 of these descriptions (i.e. nearly 55%). Interestingly, a more regular use of molecular data is observed in the description of New World Cricetidae, as compared to Old World Muridae. This probably reflects the implication of distinct working groups, each having their own procedures for the taxonomic study of these rodent families. In some cases, molecular studies identified new

Species	Family / Subfamily	Region	Main characteristics studied (diagnostic in bold)	Sources	Role of DNA sequence
Spermophilus taurensis	Sciuridae / Sciurinae	Turkey	geometric cranio-dental morphometry - karyology - DNA sequence data (cyt b, D-loop, tRNAs, X and Y chromosome sequences)	Gündüz et al. (2007)	Prominent
Heteromys catopterius	Heteromyidae / Heteromyinae	Venezuela	**external and cranio-dental morphology / morphometry**	Anderson & Gutiérrez (2009)	None
Dendromus ruppi	Nesomyidae / Dendromurinae	Sudan	**external and cranio-dental morphology / morphometry**	Dieterlen (2009)	None
Eliurus carletoni	Nesomyidae / Nesomyinae	Madagascar	**external and cranio-dental morphology / morphometry - DNA sequence data (cyt b)**	Goodman et al. (2009)	Important
Eliurus danieli	Nesomyidae / Nesomyinae	Madagascar	**external and cranio-dental morphology / morphometry**	Carleton & Goodman (2007)	Suggestive (prior to description)
Voalavo antsahabensis	Nesomyidae / Nesomyinae	Madagascar	**external and cranio-dental morphology / morphometry**	Goodman et al. (2005)	None
Proedromys liangshanensis	Cricetidae / Arvicolinae	China	**external (incl. penis) and cranio-dental morphology / morphometry**	Liu et al. (2007)	None
Peromyscus schmidlyi	Cricetidae / Neotominae	Mexico	external and cranio-dental morphology / morphometry - **karyology - DNA sequence data (cyt b)**	Bradley et al. (2004)	Important
Abrawayaomys chebezi	Cricetidae / Sigmodontinae	Argentina	**external and cranio-dental morphology / morphometry**	Pardiñas et al. (2009)	None
Akodon philipmyersi	Cricetidae / Sigmodontinae	Argentina	**external and cranio-dental morphology / morphometry - karyology - DNA sequence data (cyt b)**	Pardiñas et al. (2005)	Important
Akodon polopi	Cricetidae / Sigmodontinae	Argentina	**external and cranio-dental morphology / morphometry - DNA sequence data (cyt b)**	Jayat et al. (2010)	Important
Akodon viridescens	Cricetidae / Sigmodontinae	Argentina	**external and cranio-dental morphology / morphometry - karyology - DNA sequence data (cyt b)**	Braun et al. (2010)	Prominent
Calomys cerqueirai	Cricetidae / Sigmodontinae	Brazil	**external and cranio-dental morphology / morphometry - karyology - DNA sequence data (cyt b)**	Bonvicino et al. (2010)	Important
Cerradomys goytaca	Cricetidae / Sigmodontinae	Brazil	**external and cranio-dental morphology / morphometry - karyology**	Tavares et al. (2011)	None
Cerradomys langguthi	Cricetidae / Sigmodontinae	Brazil	**external and cranio-dental morphology / morphometry - karyology - DNA sequence data (cyt b)**	Percequillo et al. (2008)	Suggestive (prior to description)
Cerradomys vivoi	Cricetidae / Sigmodontinae	Brazil	**external and cranio-dental morphology / morphometry - karyology - DNA sequence data (cyt b)**	Percequillo et al. (2008)	Suggestive (prior to description)

Drymoreomys albimaculatus	Cricetidae / Sigmodontinae	Brazil	external and cranio-dental morphology / morphometry - DNA sequence data (cyt b, IRBP)	Percequillo et al. (2011)	Important
Eligmodontia bolsonensis	Cricetidae / Sigmodontinae	Argentina	external and cranio-dental morphology / morphometry* - DNA sequence data (cyt b)	Mares et al. (2008)	Important
Juliomys ossitenuis	Cricetidae / Sigmodontinae	Brazil	external and cranio-dental morphology / morphometry - karyology - DNA sequence data (cyt b)	Costa et al. (2007)	Suggestive (prior to description)
Neusticomys ferreirai	Cricetidae / Sigmodontinae	Brazil	external and cranio-dental morphology / morphometry - karyology	Percequillo et al. (2005)	None
Oecomys sydandersoni	Cricetidae / Sigmodontinae	Bolivia	external and cranio-dental morphology / morphometry	Carleton et al. (2009)	None
Oligoryzomys moojeni	Cricetidae / Sigmodontinae	Brazil	external and cranio-dental morphology / morphometry - karyology	Weksler & Bonvicino (2005)	None
Oligoryzomys rupestris	Cricetidae / Sigmodontinae	Brazil	external and cranio-dental morphology / morphometry - karyology	Weksler & Bonvicino (2005)	None
Phyllotis alisosiensis	Cricetidae / Sigmodontinae	Argentina	external and cranio-dental morphology / morphometry - DNA sequence data (cyt b)	Ferro et al. (2010)	Important
Rhipidomys ipukensis	Cricetidae / Sigmodontinae	Brazil	external and cranio-dental morphology / morphometry - DNA sequence data (cyt b)	Rocha et al. (2011)	Important
Thomasomys andersoni	Cricetidae / Sigmodontinae	Bolivia	external and cranio-dental morphology / morphometry - karyology - DNA sequence data (cyt b)	Salazar-Bravo & Yates (2007)	Important
Lophuromys chercherensis	Muridae / Murinae	Ethiopia	external and cranio-dental morphology / morphometry - karyology - DNA sequence data (cyt b)	Lavrenchenko et al. (2007)	Important
Lophuromys kilonzoi	Muridae / Deomyinae	Tanzania	external and cranio-dental morphology / morphometry - DNA sequence data (cyt b)	Verheyen et al. (2007)	Important
Lophuromys machangui	Muridae / Deomyinae	Tanzania	external and cranio-dental morphology / morphometry - DNA sequence data (cyt b)	Verheyen et al. (2007)	Important
Lophuromys makundii	Muridae / Deomyinae	Tanzania	external and cranio-dental morphology / morphometry - DNA sequence data (cyt b)	Verheyen et al. (2007)	Important
Lophuromys menageshae	Muridae / Deomyinae	Ethiopia	external and cranio-dental morphology / morphometry - karyology - DNA sequence data (cyt b)	Lavrenchenko et al. (2007)	Important
Lophuromys pseudosikapusi	Muridae / Murinae	Ethiopia	external and cranio-dental morphology / morphometry - karyology - DNA sequence data (cyt b)	Lavrenchenko et al. (2007)	Important
Lophuromys sabunii	Muridae / Deomyinae	Tanzania	external and cranio-dental morphology / morphometry - DNA sequence data (cyt b)	Verheyen et al. (2007)	Important

Species	Family / Subfamily	Locality	Data types	Reference	Importance
Lophuromys stanleyi	Muridae / Deomyinae	Tanzania	external and cranio-dental morphology / morphometry - DNA sequence data (cyt b)	Verheyen et al. (2007)	Important
Archboldomys kalinga	Muridae / Murinae	Philippines	external and cranio-dental morphology / morphometry - karyology	Balete et al. (2006)	None
Chrotomys sibuyanensis	Muridae / Murinae	Philippines	external and cranio-dental morphology / morphometry - DNA sequence data (cyt b)	Rickart et al. (2005)	Important
Coccymys kirrhos	Muridae / Murinae	New Guinea	external and cranio-dental morphology / morphometry	Musser & Lunde (2009)	None
Grammomys brevirostris	Muridae / Murinae	Kenya	external and cranio-dental morphology / morphometry	Kryštufek (2008)	None
Grammomys selousi	Muridae / Murinae	Tanzania	external and cranio-dental morphology / morphometry - karyology	Denys et al. (2011)	None
Hydromys ziegleri	Muridae / Murinae	New Guinea	external and cranio-dental morphology / morphometry	Helgen (2005b)	None
Hylomyscus arcimontensis	Muridae / Murinae	Tanzania	external and cranio-dental morphology / morphometry	Carleton & Stanley (2005)	None
Hylomyscus walterverheyeni	Muridae / Murinae	Gabon	external and cranio-dental morphology / morphometry - karyology - DNA sequence data (cyt b, 16S rRNA)	Nicolas et al. (2008)	Suggestive (prior to description)
Leptomys arfakensis	Muridae / Murinae	New Guinea	external and cranio-dental morphology / morphometry	Musser et al. (2008)	None
Leptomys paulus	Muridae / Murinae	New Guinea	external and cranio-dental morphology / morphometry	Musser et al. (2008)	None
Mayermys germani	Muridae / Murinae	New Guinea	external and cranio-dental morphology / morphometry	Helgen (2005a)	None
Microhydromys argenteus	Muridae / Murinae	New Guinea	external and cranio-dental morphology / morphometry	Helgen et al. (2010)	None
Mus cypriacus	Muridae / Murinae	Cyprus	classical and geometric external, cranio-dental and dental morphometry - karyology - DNA sequence data (D-loop) and other molecular markers	Cucchi et al. (2006)	Suggestive (prior to description)
Musseromys gulantang	Muridae / Murinae	Philippines	external and cranio-dental morphology / morphometry - DNA sequence data (GHR & IRBP)	Heaney et al. (2009)	Important
Rhynchomys banahao	Muridae / Murinae	Philippines	external and cranio-dental morphology / morphometry	Balete et al. (2007)	None

Rhynchomys tapulao	Muridae / Murinae	Philippines	external and cranio-dental morphology / morphometry	Balete et al. (2007)	None
Saxatilomys paulinae	Muridae / Murinae	Laos	external and cranio-dental morphology / morphometry	Musser et al. (2005)	None
Tonkinomys daovantieni	Muridae / Murinae	Laos	external and cranio-dental morphology / morphometry	Musser et al. (2006)	None
Laonastes aenigmamus	Diatomyidae	Laos	external and cranio-dental morphology / morphometry - DNA sequence data (cyt b & 12S rRNA)	Jenkins et al. (2005)	Important
Phyllomys sulinus	Echymyidae	Brazil	external and cranio-dental morphology / morphometry - karyology	Leite et al. (2008)	Suggestive (prior to description)
Isothrix barbarabrownae	Echymyidae	Peru	external and cranio-dental morphology / morphometry	Patterson & Velazco (2006)	None

Table 1. Rodent species descriptions in the period 2004-2011, and role of DNA sequence data among the datasets used.

lineages that were subsequently formally described as new species (these descriptions then included or not the sequence data). In most cases, the description process encompassed the DNA sequence data that sometimes represented the major characteristics of the newly described species. Indeed, both Gündüz et al. (2007) and Braun et al. (2010) when describing *Spermophilus taurensis* and *Akodon viridescens*, respectively, pointed out the difficulty of assigning individual specimens to these species based on the sole morphological and morphometrical characteristics, whereas molecular data unambiguously classify the same specimens in well-supported monophyletic clades. To a lesser extent, this is the case for the numerous species of *Lophuromys* described by Lavrenchenko et al. (2007) and Verheyen et al. (2007; Table 1). In the latter paper however, some of these new species would certainly need a stronger assessment of their status, as their current assignation relied on insufficiently supported bootstrap values in distance-based phylogenetic trees built on cytochrome b sequences (see for instance graph 10 in Verheyen et al., 2007).

In all studies using DNA sequence data but three (Cucchi et al., 2006, for *Mus cypriacus*, Heaney et al., 2009 for *Musseromys gulantang*, and Jenkins et al., 2005 for *Laonastes aenigmamus*), the molecular analyses were conducted using the cytochrome b gene. This proves again the importance of this gene at the species level in rodents (and more generally in mammals), and highlights its potential importance as a target gene for DNA barcoding / DNA taxonomy in this group (see above). Cytochrome b was regularly considered in association with other, mainly mitochondrial, genes. Karyological data were included in 21 of the 51 recent species descriptions, but proved to be not systematically diagnostic. Around 2/3 of these recent species descriptions were based on a combination of morphologic / morphometric and genetic (*sensu lato*) data that were jointly used to delimit and characterize the new species. This again testifies for the importance of adopting an integrative approach in modern taxonomy (Dayrat, 2005; Lecompte et al., 2003), in which molecular data will undoubtedly have a growing place in conjunction with other fields including traditional ones.

5. Conclusion

It is now clear that classical morpho-anatomical data will not alone allow answer the numerous questions, and test the numerous hypotheses that remain in the field of animal systematics, even in structurally complex groups such as higher Vertebrates. Along the years, DNA sequence data have proven to represent a precious alternative source of information, as exemplified here above at different taxonomic levels in Rodentia. At the genus level, which has not been tackled in details here above, one may find additional examples acknowledging the same fact: genera defined on morphological grounds only often need to be revised using complementary tools, and particularly molecular ones. This aspect has been underlined at various occasions by Musser & Carleton (2005; see for instance their comments on the relatively recently described genera *Amphinectomys*, *Andalgalomys* and *Volemys*). At this important taxonomic level, the use of molecular data should also be encouraged, even if other sources of information may usefully be considered (see Ford [2006] for an interesting discussion centred on Australian murid rodents).

This overview of the role that molecular data have played in rodent systematics over the last decades reflects the ever growing importance of this kind of information in evolutionary biology as a whole. The improvement of laboratory procedures, the development of sophisticated data treatment softwares and of huge databases, together with the evolution of concepts associated with this field of research, have concurred to make DNA sequences a major source of information for disciplines such as population genetics, phylogeography and phylogeny, that all are related to some extent with systematics. From there, additionally to the classic work they need to do on voucher specimens, taxonomists now have also to get involved in various activities, or at least to consider the results from several disciplines when revising a group or describing a new species. These activities will undoubtedly continue to include the acquisition, treatment and interpretation of molecular data. In rodents and as shown above, a number of taxa (from species to higher-level supraspecific groups) are still in need of a more accurate delimitation, phylogenetic relationships of many others still remain to be established, and new species still await to be described all over the world!

6. Acknowledgments

We warmly thank Pascale Chevret and Jacques Michaux for their comments and useful suggestions on a previous draft of this text.

7. References

Abramson, N.I., Lebedev, V.S., Tesakov, A.S. & Bannikova, A.A. (2009). Supraspecies relationships in the subfamily Arvicolinae (Rodentia, Cricetidae): An unexpected result of nuclear gene analysis. *Molecular Biology*, 43, 5, 834–846.

Acosta, M.J., Marchal, J.A., Fernandez-Espartero, C., Romero-Fernandez, I., Rovatsos, M.T., Giagia-Athanasopoulou, E.B., Gornung, E., Castiglia, R., & Sanchez, A. (2010). Characterization of the satellite DNA Msat-160 from species of *Terricola* (*Microtus*) and *Arvicola* (Rodentia, Arvicolinae). *Genetica*, 138, 9-10,1085–1098.

Amrine-Madsen, H., Koepfli, K.P., Wayne, R.K. & Springer, M.S. (2003). A new phylogenetic marker, apolipoprotein B, provides compelling evidence for eutherian relationships. *Molecular Phylogenetics and Evolution*, 28, 2, 225-240.

Anderson, R. P. & Gutiérrez, E. E. (2009). Taxonomy, distribution, and natural history of the genus *Heteromys* (Rodentia: Heteromyidae) in central and eastern Venezuela, with the description of a new species from the Cordillera de la Costa. In R. S. Voss and M. D. Carleton (editors), Systematic mammalogy: contributions in honor of Guy G. Musser. Bulletin of the American Museum of Natural History 331: 33–93.

Anderson, S. & Jones, J.K. (1984). *Orders and families of recent mammals of the world*, Anderson, S. & Jones, Jr. J. K., John Wiley and Sons, New York.

Arnason, U., Adegoke, J.A., Bodin, K., Born, E.W., Esa, Y.B., Gullberg, A., Nilsson, M., Short, R.V., Xu, X. & Janke, A. (2002). Mammalian mitogenomic relationships and the root of the eutherian tree. *Proceedings of the National Academy of Sciences of the United States of America*, 99, 12, 8151-8156.

Avise JC. 2000. Phylogeography: the history and formation of species. Harvard University press: Cambridge.

Avise, J.C. & Aquadro, C.F. (1982). A comparative summary of genetic distances in the vertebrates. *Evolutionary Biology*, 15, 151–184.

Balete, D.S., Rickart, E.A. & Heaney, L.R. (2006). A new species of the shrew-mouse, *Archboldomys* (Rodentia: Muridae: Murinae), from the Philippines. *Systematics and Biodiversity*, 4, 4, 489–501.

Balete, D.S., Rickart, E.A., Rosell-Ambal, R.G.B., Jansa, S. & Heaney, L.R. (2007). Descriptions of two new species of *Rhynchomys* Thomas (Rodentia : Muridae : Murinae) from Luzon Island, Philippines. *Journal of Mammalogy*, 88, 2, 287-301.

Baker, R.J. & Bradley, R.D. (2006). Speciation in mammals and the genetic species concept. *Journal of Mammalogy*, 87, 4, 643–662.

Bickford, D., Lohman, D.J., Sodhi, N.S., Ng, P.K.L., Meier, R., Winker, K., Ingram, K.K. & Das, I. (2007) Cryptic species as a window on diversity and conservation. *Trends in Ecology & Evolution*, 22, 3, 148-155.

Blanga-Kanfi, S., Miranda, H., Penn, O., Pupko, T., DeBry, R. W. & Huchon, D. (2009). Rodent phylogeny revised: analysis of six nuclear genes from all major rodent clades. *BMC Evolutionary Biology*, 9, 71.

Blaxter, M. L. (2003). Molecular systematics: counting angels with DNA. *Nature* 421, 122–124.

Blaxter, M. L. (2004). The promise of a DNA taxonomy. *Philosophical Transactions of the Royal Society, London, B*, 359, 1444, 669-679.

Bonvicino, C.R., De Oliveira, J.A. & Gentile, R. (2010). A new species of *Calomys* (Rodentia: Sigmodontinae) from Eastern Brazil. *Zootaxa* 2336, 19–25.

Bradley, R.D. & Baker, R.J. (2001). A test of the genetic species concept: Cytochrome b sequences and mammals. *Journal of Mammalogy*, 82, 4, 960–973.

Bradley, R.D., Carroll, D.S., Haynie, M.L., Muñiz Martínez, R., Hamilton, M.J. & Kilpatrick, C.W. (2004). A new species of *Peromyscus* from Western Mexico. *Journal of Mammalogy*, 85, 6, 1184–1193.

Bradley, R.D., Edwards, C.W., Carroll, D.S. & Kilpatrick, C.W. (2004). Phylogenetic relationships of Neotomine – Peromyscine rodents: Based on DNA sequences from the mitochondrial cytochrome-b gene. *Journal of Mammalogy*, 85, 3, 389-395.

Brandt, J. F. (1855). Beiträge zur nähern Kenntniss der Säugethiere Russlands. *Mémoires de l'Académie Impériale des Sciences de St. Pétersbourg*, 6–9, 1–375.

Braun, J.K., Mares, M.A., Coyner, B.S. & Van den Bussche, R.A. (2010). New species of *Akodon* (Rodentia: Cricetidae: Sigmodontinae) from central Argentina. *Journal of Mammalogy*, 91, 2, 387–400.

Bryja, J., Granjon, L., Dobigny, G., Patzenhauerová, H., Konečný, A., Duplantier, J.M., Gauthier, P., Colyn, M., Durnez, L., Lalis, A. & Nicolas, V., 2010 - Plio-Pleistocene history of West African Sudanian savanna and the phylogeography of the *Praomys daltoni* complex (Rodentia): The environment / geography / genetic interplay. *Molecular Ecology* 19, 21, 4783–4799.

Butler, P.M. (1985). Homologies of molar cusps and crests, and their bearing on assessments of rodent phylogeny. In: *Evolutionary Relationships Among Rodents : a Multidisciplinary Analysis*, Luckett, W.P. & Hartenberger, J.-L., pp. 381-402, Plenum press, New York.

Bugge, J. (1985). Systematic value of the carotid arterial pattern in rodents. In: *Evolutionary Relationships Among Rodents : a Multidisciplinary Analysis*, Luckett, W.P. & Hartenberger, J.-L., pp. 355-380, Plenum press, New York.

Cao, Y., Adachi, J. & Hasegawa, M. (1998). Comment on the quartet puzzling method for finding maximum-likelihood tree topologies. *Molecular Biology and Evolution*, 15, 1, 87-89.

Carleton, M. D. (1984). Introduction to rodents. In: *Orders and Families of Recent Mammals of the World*, Anderson, S. & Jones, J. K., pp. 255–288, Wiley, New-York.

Carleton, M.D. & Musser, G.G. (2005). Order Rodentia. In: *Mammal Species of the World: A Taxonomic and Geographic reference*, Wilson, D.E. & Reeder, D.M., pp. 745-752, Johns Hopkins University Press, Baltimore.

Carleton, M.D. & Stanley, W.T. (2005). Review of the *Hylomyscus denniae* complex (Rodentia: Muridae) in Tanzania, with a description of a new species. *Proceedings of the Biological Society of Washington*, 118, 3, 619-646.

Carleton, M.D. & Goodman, S.M. (2007). A new species of the *Eliurus majori* complex (Rodentia: Muroidea: Nesomyidae) from South-central Madagascar, with remarks on emergent species groupings in the genus *Eliurus*. *American Museum Novitates*, 3547, 1-21.

Carleton, M.D., Emmons, L.H. & Musser, G.G. (2009). A new species of the Rodent genus *Oecomys* (Cricetidae: Sigmodontinae: Oryzomyini) from Eastern Bolivia, with emended definitions of *O. concolor* (Wagner) and *O. mamorae* (Thomas). *American Museum Novitates*, 3661, 1-32.

Catzeflis, F.M., Aguilar, J.P. & Jaeger, J.J. (1992). Muroid Rodents – Phylogeny and evolution. *Trends in Ecology and Evolution*, 7, 4, 122-126.

Chevret, P., Denys, C., Jaeger, J. J., Michaux, J. & Catzeflis, F. M. (1993). Molecular evidence that the spiny mouse (*Acomys*) is more closely related to gerbils (Gerbillinae) than to true mice (Murinae). *Proceedings of the National Academy of Sciences USA*, 90, 8, 3433- 3436.

Chevret, P. & Dobigny, G (2005). Systematics and evolution of the subfamily Gerbillinae (Mammalia, Rodentia, Muridae). *Molecular Phylogenetics and Evolution*, 35, 3, 674-688.

Churakov, G., Sadasivuni, M.K., Rosenbloom, K.R., Huchon, D., Brosius, J. & Schmitz, J. (2010). Rodent Evolution: Back to the Root. *Molecular Biology and Evolution*, 27, 6, 1315-1326.

Coleman, A.W. (2009). Is there a molecular key to the level of "biological species" in eukaryotes? A DNA guide. *Molecular Phylogenetics and Evolution*, 50, 1, 197-203.

Conroy, C.J. & Cook, J.A. (1999). MtDNA evidence for repeated pulses of speciation within arvicoline and murid rodents. *Journal of Mammalian Evolution*, 6, 3, 221-245.

Costa, L.P., Pavani, S.E., Leite, Y.L.R. & Fagundes, V. (2007). A new species of *Juliomys* (Mammalia: Rodentia: Cricetidae) from the Atlantic forest of southeastern Brazil. *Zootaxa*, 1463, 21–37.

Cucchi, T., Orth, A., Auffray, J.C., Renaud, S., Fabres, L., Catalan, J., Hadjisterokotis, E., Bonhomme, F. & Vigne, J.D. (2006). A new endemic species of the subgenus *Mus* (Rodentia, Mammalia) on the Island of Cyprus. *Zootaxa*, 1241, 1–36.

Dayrat, B. (2005). Towards integrative taxonomy. *Biological Journal of the Linnean Society*, 85, 3, 407–415.

DeBry, R. W., and Sagel, R. M. (2001). Phylogeny of Rodentia (Mammalia) inferred from the nuclear-encoded gene IRBP. *Molecular Phylogenetic and Evolution*, 19, 2, 290–301.

DeBry, R. W. (2003). Identifying conflicting signal in a multigene analysis reveals a highly resolved tree: the phylogeny of Rodentia (Mammalia). *Systematic Biology*, 52, 5, 604-617.

D'Elia, G., Gonzalez, E.M. & Pardinas, U.F.J. (2003). Phylogenetic analysis of sigmodontine rodents (Muroidea), with special reference to the akodont genus *Deltamys*. *Mammalian Biology*, 68, 6, 351-364.

Delsuc, F., Brinkmann, H. & Philippe, H. (2005). Phylogenomics and the reconstruction of the tree of life. *Nature Reviews Genetics*, 6, 361-375.

Denys, C., Michaux, J., Catzeflis, F., Ducrocq, S. & Chevret, P. (1995). Morphological and molecular data against the monophyly of the Dendromurinae (Rodentia : Muridae). *Bonner Zoologische Beiträge*, 45, 3-4, 173-190.

Denys, C., Lalis, A., Lecompte, E., Cornette, R., Moulin, S., Makundi, R.H., Machang'u, R.S., Volobouev, V. & Aniskine, V.M. (2011). A faunal survey in Kingu Pira (south Tanzania), with new karyotypes of several small mammals and the description of a new Murid species (Mammalia, Rodentia) *Zoosystema*, 33, 1, 5-47.

De Queiroz, K. (2007). Species concepts and species delimitation. *Systematic Biology*, 56, 6, 879–886.

De Queiroz, K. (2011). Branches in the lines of descent: Charles Darwin and the evolution of the species concept. *Biological Journal of the Linnean Society*, 103, 1, 19–35.

D'Erchia, A.M., Gissi, C., Pesole, G., Saccone, C. & Arnason U. (1996). The guinea-pig is not a rodent. *Nature*, 381, 597-600.

Dieterlen, F. (2009). Climbing mice of the genus *Dendromus* (Nesomyidae, Dendromurinae) in Sudan and Ethiopia, with the description of a new species. *Bonner zoologische Beiträge*, 56, 3, 185-200.

Dobigny, G., Granjon, L., Aniskin, V., Bâ, K., & Volobouev, V. (2003). A new sibling species of *Taterillus* (Muridae, Gerbillinae) from West Africa. *Mammalian Biology*, 68, 5, 299-316.

Dobzhansky, T. (1950). Mendelian populations and their evolution. *The American Naturalist*, 84, 819, 401-418.

Dubois, J.Y., Rakotondravony, D., Hänni, C., Sourrouille, P. & Catzeflis, F.M. (1996). Molecular evolutionary relationships of three genera of Nesomyinae, endemic rodent taxa from Madagascar. *Journal of Mammalian Evolution*, 3, 3, 239-260.

Dubois, J.Y.F., Catzeflis, F.M. & Beintema, J.J. (1999). The phylogenetic position of "Acomyinae" (Rodentia, Mammalia) as sister group of a Murinae + Gerbillinae clade: Evidence from the nuclear ribonuclease gene. *Molecular Phylogenetics and Evolution*, 13, 1, 181-192.

Ducroz, J.F., Volobouev, V. & Granjon, L. (2001). An assessement of systematics of Arvicanthine rodents using mitochondrial DNA sequences : Evolutionary and biogeographical implications. *Journal of Mammalian Evolution*, 8, 3, 173-206.

Engel, S.R., Hogan, K.M., Taylor, J.F. & Davis, S.K. (1998). Molecular systematics and paleobiogeography of the South American rodents. *Molecular Biology and Evolution*, 15, 1, 35-49.

Ericson, P.G.P., Zuccon, D., Ohlson, J.I., Johansson, U.S., Alvarenga, H., & Prum, R.O. (2006). Higher-level phylogeny and morphological evolution of tyrant flycatchers, cotingas, manakins, and their allies (Aves : Tyrannida). *Molecular Phylogenetics and Evolution*, 40, 2, 471-483.

Fahlbusch, V. (1985). Origin and evolutionary relationships among Geomyoids. In: *Evolutionary Relationships Among Rodents : a Multidisciplinary Analysis*, Luckett, W.P. & Hartenberger, J.-L., pp. 617-630, Plenum press, New York.

Felsenstein, J. (1978). Cases in which parsimony or compatibilty methods will be positively misleading. *Systematic Zoology*, 27, 4, 401-410.

Felsenstein, J. (1985). Confidence limits on phylogenies - An approach using the bootstrap. *Evolution*, 39, 4, 783-791.

Felsenstein, J. (2004). *Inferring phylogenies*. Sinauer Associates, Inc, Sunderland, Massachussets, 664 p.

Ferguson, J.W.H. (2002). On the use of genetic divergence for identifying species. *Biological Journal of the Linnean Society*, 75, 4, 509-516.

Ferro, L.I., Martinez, J.J. & Barquez, R.M. (2010). A new species of *Phyllotis* (Rodentia, Cricetidae, Sigmodontinae) from Tucuman province, Argentina. *Mammalian Biology*, 75, 6, 523-537.

Flynn, L.J., Jacobs, L.L. & Lindsay, E.H.. (1985). Problems in Muroid phylogeny: relationship to other rodents and origin of major groups. In: *Evolutionary Relationships Among Rodents : a Multidisciplinary Analysis*, Luckett, W.P. & Hartenberger, J.-L., pp. 589-616, Plenum press, New York.

Ford, F. (2006). A splitting headache: relationships and generic boundaries among Australian murids. *Biological Journal of the Linnean Society*, 89, 1, 117-138.

Frézal, L. & Leblois, R. (2008). Four years of DNA barcoding: Current advances and prospects. *Infection, Genetics and Evolution*, 8, 5, 727-736.

Furano, A.V., Hayward, B.E., Chevret, P., Catzeflis, F. & Usdin, K. (1994). Amplification of the ancient murine Lx family of long interspersed repeated DNA occurred during the murine radiation. *Journal of Molecular Evolution*, 38, 1, 18-27.

Futuyma, D.J. (1998). *Evolutionary biology*, 3rd edn. Sinauer: Sunderland, Massachusetts.

Giribet G., Dunn, C.W., Edgecombe, G.D. & Rouse, G.W. (2007). A modern look at the Animal Tree of Life. *Zootaxa*, 1668, 61-79.

Gogolevskaya, I.K., Veniaminova, N.A. & Kramerov, D.A. (2010). Nucleotide sequences of B1 SINE and 4.5SI RNA support a close relationship of zokors to blind mole rats (Spalacinae) and bamboo rats (Rhizomyinae). *Gene*, 460, 1-2, 30-38.

Goodman, S.M., Rakotondravony, D., Randriamanantsoa, H.N. & Rakotomalala-Razanahoera, M. (2005). A new species of rodent from the montane forest of central eastern Madagascar (Muridae: Nesomyinae: *Voalavo*). *Proceedings of the Biological Society of Washington*, 118, 4, 863-873.

Goodman, S.M., Raheriarisena, M. & Jansa, S.A. (2009). A new species of *Eliurus* Milne Edwards, 1885 (Rodentia: Nesomyinae) from the Réserve Spéciale d'Ankarana, northern Madagascar. *Bonner zoologische Beiträge*, 56, 3, 133–149.

Graf, J. D. (1982). Génétique biochimique, zoogéographie et taxonomie des Arvicolidae (Mammalia, Rodentia). *Revue Suisse de Zoologie*, 89, 3, 749-787.

Granjon, L. & Dobigny, G. (2003). The importance of chromosomally-based identifications for correct understanding of African rodent zoogeography: Lake Chad murids as an example. *Mammal Review*, 33, 1, 77-91.

Graur, D., Hide, W.A. & Li, W.-H. (1991). Is the guinea-pig a rodent? *Nature*, 351: 649-652.

Gregory, W.K. (1910). *The orders of mammals. Bulletin of the American Museum of Natural History*, 27, 1-524.

Gündüz, I., Jaarola, M., Tez, C., Yeniyurt, C., Polly, P.D. & Searle, J.B. (2007). Multigenic and morphometric differentiation of ground squirrels (*Spermophilus*, Scuiridae, Rodentia) in Turkey, with a description of a new species. *Molecular Phylogenetics and Evolution*, 43, 3, 916–935.

Hajibabaei, M., Singer, G.A.C., Hebert, P.D.N. & Hickey, D.A. (2007). DNA barcoding: how it complements taxonomy, molecular phylogenetics and population genetics, *Trends in Genetics*, 23, 4, 167-172.

Harrison, R. G. (1998). Linking evolutionary pattern and process. Pages 19–31 in Endless forms: Species and speciation (D. J. Howard, and S. H. Berlocher, eds.). Oxford University Press, New York.

Harshman, J., Braun, E.L., Braun, M.J., Huddleston, C. J., Bowie, R.C.K., Chojnowskid, J.L., Hackett, S.J., Han, K.-H et al. (2008). Phylogenomic evidence for multiple losses of flight in ratite birds. *Proceedings of the National Academy of Sciences of the United States of America*, 105, 36, 13462-13467.

Hartenberger, J.-L. (1985). The order Rodentia : Major questions on their evolutionary origin, relationships and suprafamily systematics. In: *Evolutionary Relationships Among Rodents: a Multidisciplinary Analysis*, Luckett, W.P. & Hartenberger, J.-L., pp. 1-33, Plenum Press, New York.

Hartenberger, J.-L. (1996). Les débuts de la radiation adaptative des Rodentia (Mammalia). *Comptes Rendus de l'Académie des Sciences*, 323, 7, 631-637.

Hartenberger, J.-L. (1998). Description de la radiation des Rodentia (Mammalia) du Paléocène supérieur au Miocène; incidences phylogénétiques. *Comptes Rendus de l'Académie des Sciences*, 326, 6, 439-444.

Hautier, L., Lebrun, R., Saksiri, S., Michaux, J., Vianey-Liaud, M. & Marivaux, L. (2011). Hystricognathy *vs* Sciurognathy in the rodent jaw: A new morphometric assessment of Hystricognathy applied to the living fossil *Laonastes* (Diatomyidae). *PLoS ONE* 6, 4.

Heaney, L.R., Balete, D.S., Rickart, E.A. Josefa Veluz, M. & Jansa, S.A. (2009). A new genus and species of small 'Tree-Mouse' (Rodentia, Muridae) related to the Philippine Giant Cloud Rats. *Bulletin of the American Museum of Natural History*, 331, 1, 205-229.

Hebert, P.D.N., Cywinska, A., Ball, S.L. & deWaard, J.R. (2003). Biological identifications through DNA barcodes. *Proceedings of the Royal Society of London, B*, 270, 1512, 313-321.

Helgen, K.M. (2005a). A new species of murid rodent (genus *Mayermys*) from south-eastern New Guinea. *Mammalian Biology*, 6, 1, 61-67.

Helgen, K.M. (2005b). The amphibious murines of New Guinea (Rodentia, Muridae): the generic status of *Baiyankamys* and description of a new species of *Hydromys*. *Zootaxa*, 913, 1-20.

Helgen, K.M., Leary, T. & Aplin, K.P. (2010). A review of *Microhydromys* (Rodentia: Murinae), with description of a new species from southern New Guinea. *American Museum Novitates*, 3676, 1-22.

Hoffmann, M., Grubb, P., Groves, C.P., Hutterer, R., Van der Straeten, E., Simmons, N. & Bergmans, W. (2009) A synthesis of African and western Indian Ocean Island mammal taxa (Class: Mammalia) described between 1988 and 2008: an update to Allen (1939) and Ansell (1989). *Zootaxa*, 2205, 1-36.

Honeycutt, R.L. & Adkins, R.M. (1993). Higher level systematics of eutherian mammals: an assessment of molecular characters and phylogenetic hypotheses. *Annual Review of Ecology and Systematics*, 24: 279-305.

Horvath, J.E., Weisrock, D.W., Embry, S.L., Fiorentino, I., Balhoff, J.P., Kappeler, P., et al. (2008). Development and application of a phylogenomic toolkit: Resolving the evolutionary history of Madagascar's lemurs. *Genome Research*, 18, 3, 489-499.

Huchon, D., Catzeflis, F. M. & Douzery, E. J. P. (1999). Molecular evolution of the nuclear von Willebrand factor gene in mammals and the phylogeny of rodents. *Molecular Biology and Evolution*, 16, 5, 577-589.

Huchon, D., Madsen, O., Sibbald, M.J.J.B., Ament, K., Stanhope, M.J., Catzeflis, F., de Jong, W.W., Douzery, E.J.P. (2002). Rodent phylogeny and a timescale for the evolution of Glires: evidence from an extensive taxon sampling using three nuclear genes. *Molecular Biology and Evolution*, 19, 7, 1053-1065.

Huchon, D., Chevret, P., Jordan, U., Kilpatrick, C.W., Ranwez, V., Jenkins, P.D. *et al.* (2007). Multiple molecular evidences for a living mammalian fossil. *Proceedings of the National Academy of Sciences of the United States of America*, 104, 18, 7495-7499.

Jaeger, J.-J. (1988) Rodent phylogeny: New data and old problems. In: *The Phylogeny and Classification of the Tetrapods*, Benton, M.J., pp. 177-199, Clarendon, Oxford.

Jansa, S.A., Goodman, S.M. & Tucker, P.K. (1999). Molecular phylogeny and biogeography of the native rodents of Madagascar (Muridae: Nesomyinae): A test of the single-origin hypothesis. *Cladistics*, 15, 3, 253-270.

Jansa, S.A., Carleton, M.D. (2003). Systematics and phylogenetics of Madagascar's native rodents, In: *The Natural History of Madagascar*, Goodman, S.M. & Benstead, J.P. (Eds.), pp. 1257–1265, The University of Chicago Press, Chicago.

Jansa, S.A. & Weksler, M. (2004). Phylogeny of muroid rodents: relationships within and among major lineages as determined by IRBP gene sequences. *Molecular Phylogenetics and Evolution*, 31, 1, 256-276.

Jansa, S.A., Barker, F.K. & Heaney, L.R. (2006). The pattern and timing of diversification of Philippine endemic rodents: evidence from mitochondrial and nuclear gene sequences. *Systematic Biology*, 55,1, 73-88.

Jansa, S.A., Giarla, T.C. & Lim, B.K. (2009). The Phylogenetic position of the rodent genus *Typhlomys* and the geographic origin of Muroidea. *Journal of Mammalogy*, 90, 5, 1083-1094.

Jayat, J.P., Ortiz, P.E., Salazar-Bravo, J., Pardiñas, U.F.J. & D'Elia, G. (2010). The *Akodon boliviensis* species group (Rodentia: Cricetidae: Sigmodontinae) in Argentina: species limits and distribution, with the description of a new entity. *Zootaxa*, 2409, 1–61.

Jeffroy, O., Brinkmann, H., Delsuc, F. & Philippe, H. (2006). Phylogenomics: the beginning of incongruence? *Trends in Genetics*, 22, 4, 225-231.

Jenkins, P.D., Kilpatrick, C.W., Robinson, M.F. & Timmins, R.J. (2005). Morphological and molecular investigations of a new family, genus and species of rodent (Mammalia : Rodentia : Hystricognatha) from Lao PDR. *Systematics and Biodiversity*, 2, 4, 419-454.

Johns, G. C. & Avise, J.C. (1998). A comparative summary of genetic distances in the vertebrates from the mitochondrial cytochrome b gene. *Molecular Biology and Evolution*, 15, 11, 1481–1490.

Kjer, K.M. & Honeycutt, R.L. (2007). Site specific rates of mitochondrial genomes and the phylogeny of eutheria. *BMC Evolutionary Biology*, 7, 8.

Knowlton, N. (1986). Cryptic species and sibling species among the decapod Crustacea. *Journal of Crustacean Biology*, 6, 3, 356-363.

Kryštufek, B. (2008). Description of a new thicket rat from Kenya: *Grammomys brevirostris* n. sp. *Acta Zoologica Lituanica*, 18, 4, 221 -227.

Lavocat, R. (1973). Les rongeurs du Miocène d'Afrique Orientale. *Mémoires et Travaux de l'Ecole Pratique des Hautes Etudes, Montpellier*, 1, 1-284.

Lavocat, R. (1978). Rodentia and Lagomorpha, In : *Evolution of African mammals*, Maglio, V.J. & Cooke, H.B.S. (Eds), pp. 69-89, Harvard University Press, Cambridge.

Lavocat, R. & Parent, J.-P. (1985). Phylogenetic analysis of middle ear features in fossil and living rodents. In: *Evolutionary Relationships Among Rodents : a Multidisciplinary Analysis*, Luckett, W.P. & Hartenberger, J.-L., pp. 333-354, Plenum press, New York.

Lavrenchenko, L.A., Verheyen, W.N., Verheyen, E., Hulselmans, J. & Leirs, H. (2007). Morphometric and genetic study of Ethiopian *Lophuromys flavopunctatus* Thomas, 1888 species complex with description of three new 70-chromosomal species (Muridae, Rodentia). *Bulletin de l'Institut Royal des Sciences Naturelles de Belgique, Biologie*, 77, 77-117.

Lazzari, V., Charles, C., Tafforeau, P., Vianey-Liaud, M., Aguilar, J.-P., Jaeger, J.-J., Michaux, J. & Viriot, L. (2008) Mosaic Convergence of Rodent Dentitions. *PLoS ONE* 3, 10.

Lecointre, G., Philippe, H., Lê, H.L.V. & Le Guyader, H. (1993). Species sampling has a major impact on phylogenetic inference. *Molecular Phylogenetics and Evolution*, 2, 3, 205-224.

Lecointre, G., Le Guyader, H., Visset, D. & McCoy, K. (2006). *The tree of life: a phylogenetic classification*, Belknap Press of Harvard, University Cambridge, Massachusetts.

Lecompte, E., Granjon, L., Peterhans, J.K. & Denys, C. (2002). Cytochrome b-based phylogeny of the *Praomys* group (Rodentia, Murinae): a new African radiation? *Comptes Rendus Biologies* 325, 7, 827-840.

Lecompte, E., Denys, C., Granjon, L. & Volobouev, V. (2003). Integrative systematics : Contributions to *Mastomys* phylogeny and evolution. In: *Rats, mice and people: Rodent biology and management*. Singleton, G.R., Hinds, L.A., Krebs & C.J. Spratt, D.M., pp. 536-540, Australian Centre for International Agricultural Research (ACIAR), Canberra.

Lecompte, E., Denys, C. & Granjon, L. (2005). Confrontation of morphological and molecular data: The *Praomys* group (Rodentia, Murinae) as a case of adaptive convergences and morphological stasis. *Molecular Phylogenetics and Evolution*, 37, 3, 899-919.

Lecompte, E., Aplin, K., Denys C., Catzeflis, F., Chades, M. & Chevret, P. (2008). Phylogeny and biogeography of African Murinae based on mitochondrial and nuclear gene sequences, with a new tribal classification of the subfamily. *BMC Evolutionary Biology*, 8: 199.

Lee, M.S.Y., (2004). The molecularisation of taxonomy. *Invertebrate Systematics*, 18, 1, 1-6.

Leite, Y.L.R., Christoff, A.U. & Fagundes V. (2008). A new species of Atlantic forest tree rat, genus *Phyllomys* (Rodentia, Echymyidae) frou Southern Brazil. *Journal of Mammalogy*, 89, 4, 845–851.

Liu, S., Sun, Z., Zeng, Z. & Zhao, E. (2007). A New Vole (Cricetidae: Arvicolinae: *Proedromys*) from the Liangshan Mountains of Sichuan Province, China. *Journal of Mammalogy*, 88, 5, 1170-1178.

Luckett, W.P. (1977). Ontogeny of amniote foetal membranes and their application to phylogeny. In: *Major patterns in Vertebrate Evolution*, Hecht M. K., Googy, P.C. & Hecht, B. M., pp. 347-368, Plenum Press, New York.

Luckett, W.P. (1985). Superordinal and intraordinal affinities of rodents: developmental evidence from the dentition and placentation. In: *Evolutionary Relationships Among Rodents : a Multidisciplinary Analysis*, Luckett, W.P. & Hartenberger, J.-L., pp. 227-278, Plenum press, New York.

Luckett, W. P. & Hartenberger, J.L. (1985). *Evolutionary relationships among rodents: a multidisciplinary analysis*. Plenum press, New York, 721 p.

Luckett, W. P. & Hartenberger, J.L. (1993). Monophyly or polyphyly of the order rodentia: possible conflict between morphological and molecular interpretations. *Journal of Mammalian Evolution*, 1, 2, 127-147.

McKenna, M. C. (1961). A note on the origin of rodents. *American Museum Novitates*, 2037, 1-5.

McKenna, M.C. & Bell, S.K. (1997). *Classification of mammals above the species level*. Columbia University Press, New York, 631 p.

Maddison, W. (1989). Reconstructing Character Evolution on Polytomous Cladograms. *Cladistics*, 5, 4, 365-377.

Madsen, O., Scally, M., Douady, C., Kao, D.J., DeBry, R., Adkins, R. *et al.* (2001). Parallel adaptive radiations in two major clades of placental mammals. *Nature*, 409, 610-614.

Mares, M.A., Braun, J.K., Coyner, B.S. & Van den Bussche, R.A. (2008). Phylogenetic and biogeographic relationships of gerbil mice *Eligmodontia* (Rodentia, Cricetidae) in South America, with a description of a new species. *Zootaxa*, 1753, 1-33.

Mayr, E. (1948). The bearing of the new systematics on genetical problems. The nature of species. *Advances in Genetics*, 2, 205-237.

Marivaux, L., Vianey-Liaud, M. & Jaeger, J.-J. (2004). High-level phylogeny of early Tertiary rodents: dental evidence. *Zoological Journal of the Linnean Society*, 142, 1, 105-134.

Martin, T. (1997). Incisor enamel microstructure and systematics in rodents. In: *Tooth enamel microstructure*, von Koenigswald, W. & Sander, PM, pp. 163-175, Balkema, Rotterdam.

Michaux, J. & Catzeflis, F. (2000). : The bushlike radiation of muroid rodents is exemplified by the molecular phylogeny of the LCAT nuclear gene. *Molecular Phylogenetics and Evolution*, 17, 2, 280-293.

Michaux, J.R., Reyes, A. & Catzeflis, F. (2001). Evolutionary history of the most speciose mammals: molecular phylogeny of muroid rodents. *Molecular Biology and Evolution*, 18, 11, 2017-2031.

Miller, J.R. & Engstrom, M.D. (2008). The relationships of major lineages within Peromyscine rodents: A molecular phylogenetic hypothesis and systematic reappraisal. *Journal of Mammalogy*, 89, 5, 1279-1295.

Montgelard, C., Bentz, S., Tirard, C., Verneau, O. & Catzeflis, F.M. (2002). Molecular systematics of Sciurognathi (Rodentia): the mitochondrial cytochrome b and 12S rRNA genes support the Anomaluroidea (Pedetidae and Anomaluridae). *Molecular Phylogenetics and Evolution*, 22, 2, 220-233.

Montgelard, C., Forty, E., Arnal, V. & Matthee, C.A. (2008). Suprafamilial relationships among Rodentia and the phylogenetic effect of removing fast-evolving nucleotides in mitochondrial, exon and intron fragments. *BMC Evolutionary Biology*, 8, 321.

Murphy, W.J., Eizirik, E., Johnson, W.E., Zhang, Y.P., Ryder, O.A. & O'Brien, S.J. (2001). Molecular phylogenetics and the origins of placental mammals. *Nature*, 409, 614-618.

Musser, G.G. & Carleton, M.D. (2005). Superfamily Muroidea, In: *Mammal species of the world A taxonomic and geographic reference, Volume 2*, Wilson, D.E. & Reeder, D.M. (Eds), pp. 894-1531, Johns Hopkins University, Baltimore.

Musser, G.G. & Lunde D.P. (2009). Systematic reviews of New Guinea *Coccymys* and *"Melomys"* albidens (Muridae, Murinae) with descriptions of new taxa. *Bulletin of the American Museum of Natural History*, 329, 1-139.

Musser, G.G., Smith, A.L., Robinson, M.F. & Lunde, D.P. (2005). Description of a new genus and species of rodent (Murinae, Muridae, Rodentia) from the Khammouan Limestone National Biodiversity Conservation Area in Lao PDR. *American Museum Novitates*, 3497, 1-31.

Musser, G.G., Lunde, D.P. & Truong Son, N. (2006). Description of a new genus and species of rodent (Murinae, Muridae, Rodentia) from the Tower Karst region of Northeastern Vietnam. *American Museum Novitates*, 3517, 1-41.

Musser, G.G., Helgen, K.M. & Lunde, D.P. (2008). Systematic review of New Guinea *Leptomys* (Muridae, Murinae) with descriptions of two new species. *American Museum Novitates*, 3624, 1-60.

Nedbal, M. A., Honeycutt, R. L. & Schlitter, D. A. (1996). Higher-level systematics of rodents (Mammalia, Rodentia): Evidence from the mitochondrial 12S rRNA gene. *Journal of Mammalian Evolution*, 3, 3, 201–237.

Neumann, K., Michaux, J., Lebedev, V., Yigit, N., Colak, E., Ivanova, N., Poltoraus, A., Surov, A., Markov, G., Maak, S., Neumann, S. & Gattermann, R. (2006). Molecular phylogeny of the Cricetinae subfamily based on the mitochondrial cytochrome b and 12S rRNA genes and the nuclear vWF gene. *Molecular Phylogenetics and Evolution*, 39, 1, 135-148.

Nicolas, V., Wendelen, W., Barrière, P., Dudu, A. & Colyn, M. (2008). Morphometric variation in *Hylomyscus alleni* and *H. stella* (Rodentia: Muridae), and description of a new species. *Journal of Mammalogy*, 89, 1, 222–231.

Nicolas, V., Granjon, L., Duplantier, J.-M., Cruaud, A. & Dobigny, G., 2009 - Phylogeography of spiny mice (genus *Acomys*, Rodentia: Muridae), from the southwestern margin of the Sahara, with taxonomical implications. *Biological Journal of the Linnean Society*, 98, 1, 29–46.

Nishihara, H., Maruyama, S. & Okada, N. (2009). Retroposon analysis and recent geological data suggest near-simultaneous divergence of the three superorders of mammals. *Proceedings of the National Academy of Sciences of the United States of America*, 106, 13, 5235-52240.

Norris, R.W., Zhou, K., Zhou, C., Yang, G., Kilpatrick, C.W. & Honeycutt, R.L. (2004). The phylogenetic position of the zokors (Myospalacinae) and comments on the families of muroids (Rodentia). *Molecular Phylogenetics and Evolution*, 31, 3, 972–978.

Novacek, M.J. (1985). Cranial evidence for Rodent affinities. In: *Evolutionary Relationships Among Rodents : a Multidisciplinary Analysis*, Luckett, W.P. & Hartenberger, J.-L., pp. 59-82, Plenum Press, New York.

Pagès, M., Chaval, Y., Herbreteau, V., Waengsothorn, S., Cosson, J.F., Hugot, J.P., Morand, S. & Michaux, J. (2010). Revisiting the taxonomy of the Rattini tribe: a phylogeny-based delimitation of species boundaries. *BMC Evolutionary Biology*, 10, 184.

Pardiñas, U.F.J., D'Elía, G., Cirignoli, S. & Suarez, P. (2005). A new species of *Akodon* (Rodentia, Cricetidae) from the Northern Campos grasslands of Argentina. *Journal of Mammalogy*, 86, 3, 462–474.

Pardiñas, U.F.J., Teta, P. & D'Elía, G. (2009). Taxonomy and distribution of Abrawayaomys (Rodentia: Cricetidae), an Atlantic Forest endemic with the description of a new species. *Zootaxa*, 2128, 39–60.

Patterson, B.D. (2000). Patterns and trends in the discovery of new Neotropical mammals. *Diversity and Distributions*, 6, 3, 145-151.

Patterson, B.D. (2001). Fathoming tropical biodiversity: the continuing discovery of Neotropical mammals. *Diversity and Distributions*, 7, 4, 191–196.

Patterson, B.D. & Velazco, P.M. (2006). A distinctive new cloud-forest rodent (Hystricognathi: Echimyidae) from the Manu Biosphere Reserve, Peru. *Mastozoología Neotropical*, 13, 2, 175-191.

Percequillo, A.R., Carmignotto, A. P. & Silva, M. J. DE J. (2005). A new species of *Neusticomys* (Ichtyomyini, Sigmodontinae) from Central Brazilian Amazonia. *Journal of Mammalogy*, 86, 5, 873–880.

Percequillo, A.R., Hingst-Zaher, E. & Bonvicino, C.R. (2008). Systematic review of Genus *Cerradomys* Weksler, Percequillo and Voss, 2006 (Rodentia: Cricetidae: Sigmodontinae: Oryzomyini), with description of two new species from Eastern Brazil. *American Museum Novitates*, 3622, 1-46.

Percequillo, A.R., Weksler, M. & Costa, L.P. (2011). A new genus and species of rodent from the Brazilian Atlantic Forest (Rodentia: Cricetidae: Sigmodontinae: Oryzomyini), with comments on oryzomyine biogeography. *Zoological Journal of the Linnean Society*, 161, 2, 357–390.

Philippe, H. & Douzery, E. (1994). The pitfalls of molecular phylogeny based on four species as illustrated by the Cetacea/Artiodactyla relationships. *Journal of Mammalian Evolution*, 2, 2, 133-152.

Philippe, H., Delsuc, F., Brinkmann, H. & Lartillot, N. (2005). Phylogenomics. *Annual Review of Ecology, Evolution and Systematics*, 36, 541–562.

Poe, S. & Chubb, A.L. (2004). Birds in a bush: Five genes indicate explosive evolution of avian orders. *Evolution*, 58, 2, 404-415.

Pons, J., Barraclough, T. G. Gomez-Zurita, J., Cardoso, A., Duran, D. P., Hazell, S. Kamoun, S. Sumlin, W. D. & Vogler A. P. (2006). Sequence-based species delimitation for the DNA taxonomy of undescribed Insects. *Systematic Biology*, 55, 4, 595–609.

Poux, C., Madsen, O., Marquard, E., Vieites, D.R., De Jong, W.W. & Vences, M. (2005). Asynchronous colonization of Madagascar by the four endemic clades of Primates, Tenrecs, Carnivores, and Rodents as Inferred from nuclear genes. Systematic Biology, 54, 5, 719–730.

Renvoisé, E., Evans, A.R., Jebrane, A., Labruère, C., Laffont, R. & Montuire, S. (2009) Evolution of mammaltooth patterns : new insights from a developmental prediction model. *Evolution*, 63, 5, 1327-1340.

Reyes, A., Gissi C., Catzeflis, F., Nevo, E., Pesole, G., Saccone, C. (2004). Congruent mammalian trees from mitochondrial and nuclear genes using Bayesian methods. *Molecular Biology and Evolution*, 21, 2, 397-403.

Rickart, E.A., Heaney, L.R., Goodman, S.M. & Jansa, S. (2005). Review of the Philippine genera *Chrotomys* and *Celaenomys* (Murinae) and description of a new species. *Journal of Mammalogy*, 86, 2, 415-428.

Robinson, M., Catzeflis, F., Briolay, J. & Mouchiroud, D. (1997). Molecular phylogeny of rodents, with special emphasis on Murids: Evidence from nuclear gene LCAT. *Molecular Phylogenetics and Evolution*, 8, 3, 423–434.

Robovsky, J., Ricankova, V. & Zrzavy, J. (2008). Phylogeny of Arvicolinae (Mammalia, Cricetidae): utility of morphological and molecular data sets in a recently radiating clade. *Zoologica Scripta*, 37, 6, 571–590.

Rocha, R.G., Ferreira, E., Costa, B.M.A., Martins, I.C.M., Leite, Y.L.R., Costa, L.P. & Fonseca, C. (2011). Small mammals of the mid-Araguaia River in central Brazil, with the description of a new species of climbing rat. *Zootaxa* 2789, 1–34.

Rowe, K.C., Reno, M.L., Richmond, D.M., Adkins, R.M. & Steppan, S.J. (2008). Pliocene colonization and adaptive radiations in Australia and New Guinea (Sahul):

Multilocus systematics of the old endemic rodents (Muroidea: Murinae). *Molecular Phylogenetics and Evolution*, 47, 1, 84-101.

Serdobova, I.M. & Kramerov, D.A. (1998). Short retroposons of the B2 superfamily: Evolution and application for the study of rodent phylogeny. *Journal of Molecular Evolution*, 46, 2, 202-214.

Simpson, G.G. (1945). The principles of classification and a classification of mammals. *Bulletin of the American Museum of Natural History*, 85, 1-350.

Smith, M.F. & Patton, J.L. (1993). The diversification of South American murid rodents: evidence from mitochondrial DNA sequence data for the akodontine tribe. *Biological Journal of the Linnean Society*, 50, 3, 149-177.

Smith, M.F. & Patton, J.L. (1999). Phylogenetic relationships and the radiation of sigmodontine rodents in South America: Evidence from cytochrome b. *Journal of Mammalian Evolution*, 6, 2, 89-128.

Springer, M.S., DeBry, R.W., Douady, C., Amrine, H.M., Madsen, O., de Jong, W.W. & Stanhope, M.J. (2001). Mitochondrial versus nuclear gene sequences in deep-level mammalian phylogeny reconstruction. *Molecular Biology and Evolution*, 18, 2, 132-143.

Springer, M.S., Stanhope, M.J., Madsen O. & de Jong, W.W. (2004). Molecules consolidate the placental mammal tree. *Trends in Ecology and Evolution*, 19, 8, 430-438.

Steppan, S.J., Adkins, R.M. & Anderson, J. (2004). Phylogeny and divergence date estimates of rapid radiation in muroid rodents based on multiple nuclear genes. *Systematic Biology*, 53, 4, 533 -5553.

Steppan, S.J., Adkins, R.M., Spinks, P.Q. & Hale, C. (2005). Multigene phylogeny of the Old World mice, Murinae, reveals distinct geographic lineages and the declining utility of mitochondrial genes compared to nuclear genes. *Molecular Phylogenetics and Evolution*, 37, 2, 370-388.

Sullivan, J. & Swofford, D.L. (1997). Are guinea pigs rodents? The importance of adequate models in molecular phylogenetics. *Journal of Mammalian Evolution*, 4, 1, 77-86.

Tavares, W.C., Pessôa, L.M. & Gonçalves, P.R. (2011). New species of *Cerradomys* from coastal sandy plains of southeastern Brazil (Cricetidae: Sigmodontinae). *Journal of Mammalogy*, 92, 3, 645-658.

Taylor, P. (2000). Patterns of chromosomal variation in Southern African rodents. *Journal of Mammalogy*, 81, 2, 317-331.

Trouessart, E.L. (1897-1905). *Catalogus mammalium tam viventium quam fossilium. Quinquennale supplementum anno 1904*. R. Friedländer and Sohn, Berlin, 1 & 2: 1469pp. Quin. Supp.: 929pp.

Salazar-Bravo, J. & Yates, T.L. (2007). A New species of *Thomasomys* (Cricetidae: Sigmodontinae). In *The Quintessential Naturalist: Honoring the Life and Legacy of Oliver P. Pearson*, Kelt, D. A., E. P. Lessa, J. Salazar-Bravo & Patton J.L., pp. 747-774, University of California Publications in Zoology 134, Berkeley.

Usdin, K., Chevret, P., Catzeflis, F.M., Verona, R. & Furano, A.V. (1995). L1 (Line-1) retrotransposable elements provide a "fossil" record of the phylogenetic history of murid rodents. *Molecular Biology and Evolution*, 12, 1, 73-82.

Verheyen, E., Colyn, M. & Verheyen, W. (1996). A mitochondrial cytochrome *b* phylogeny confirms the paraphyly of the Dendromurinae Alston, 1896 (Muridae, Rodentia) *Mammalia*, 60, 4, 780-785.

Verheyen, W.N., Hulselmans, J.L.J., Dierckx, T., Mulungu, L., Leirs, H., Corti, M. & Verheyen, E. (2007). The characterization of the Kilimanjaro *Lophuromys aquilus* TRUE 1892 population and the description of five new *Lophuromys* species (Rodentia, Muridae). *Bulletin de l'Institut Royal des Sciences Naturelles de Belgique, Biologie*, 77, 23-75.

Vianey-Liaud, M. (1985). Possible evolutionary relationships among Eocene and lower Oligocene rodents in Asia, Europe and North America. In: *Evolutionary Relationships Among Rodents : a Multidisciplinary Analysis*, Luckett, W.P. & Hartenberger, J.-L., pp. 277-310, Plenum press, New York.

Weksler, M. (2003). Phylogeny of Neotropical oryzomyine rodents (Muridae: Sigmodontinae) based on the nuclear IRBP exon. *Molecular Phylogenetics and Evolution*, 29, 2, 331-349.

Weksler, M. & Bonvicino, C.R. (2005). Taxonomy of pigmy rice rats (genus *Oligoryzomys* (Rodentia, Sigmodontinae) of the Brazilian Cerrado, with the description of two new species. *Arquivos do Museu Nacional, Rio de Janeiro*, 63, 1, p.113-130.

Whitfield, J.B. & Lockhart, P.J. (2007). Deciphering ancient rapid radiations. *Trends in Ecology and Evolution,* 22, 5, 258-265.

Wilson, D.E. & Reeder, D.M. (1993). *Mammal species of the world: a taxonomic and geographic reference,* 2nd edn. Smithsonian Institution Press, Washington and London.

Wilson, D.E., Reeder, D.M. (2005). *Mammal Species of the World. A Taxonomic and Geographic Reference, vols. 1 and 2.* John Hopkins University Press, Baltimore.

Zimmerman, E.G., Kilpatrick, C. W. & Hart, B. J. (1978). The genetics of speciation in the rodent genus *Peromyscus. Evolution*, 32, 3, 565-579.

5

Sequencing Technologies and Their Use in Plant Biotechnology and Breeding

Victor Llaca

Dupont Agricultural Biotechnology, Pioneer Hi-Bred International,
Wilmington, Delaware,
USA

1. Introduction

The development of DNA sequencing strategies has been a high priority in genetics research since the discovery of the structure of DNA and the basic molecular mechanisms of heredity. However, it was not until the works by Maxam and Gilbert (1977), and Sanger (Sanger *et al*, 1977), that the first practical sequencing methods were developed and implemented on a large scale. The first isolation and sequencing of a plant cDNA by Bedbrook and colleagues a few years later initiated the field of Plant Molecular Genetics (Bedbrook *et al*, 1980). Plant biotechnology started shortly thereafter with the successful integration of recombinant DNA and sequencing techniques to generate the first transgenic plants using *Agrobacterium* (Fraley *et al*, 1983; Herrera-Estrella *et al*, 1983). The first genetic map in plants based on restriction fragment length polymorphisms (RFLPs; Bernatzky & Tanksley, 1986) enabled the capture of genetic variation and started the era of molecular marker-assisted plant breeding. Since then, sequencing methodologies have been essential tools in plant research. They have allowed the characterization and modification of genes and metabolic pathways, as well as the use of genetic variation for studies in species diversity, marker-assisted selection (MAS), germplasm characterization and seed purity. The determination of the reference genomes in *Arabidopsis thaliana*, rice and maize using Sanger sequencing strategies constituted major milestones that enabled the analysis of genome architecture and gene characterization in plants (The Arabidopsis Genome Initiative, 2001; International Rice Genome Project, 2005; Schnable *et al*, 2009). More recently, the development and increasing availability of multiple Next-Generation sequencing (NGS) technologies minimized research limitations and bottlenecks based on sequence information (Metzker, 2010; Glenn, 2011). It is difficult to overstate the influence that these massively parallel systems have had in our understanding of plant genomes and in the expansion, acceleration and diversification of breeding and biotechnology projects. At the same time, this influence tends to understate the importance that capillary Sanger sequencing still has in day-by-day research and development work. This review provides a description of major sequencing technologies that are available today, their use as well as future prospects in basic plant genetics research, biotechnology and breeding in crop plants.

2. Current sequencing technologies

The development of recent sequencing technologies has generated a remarkable increase, by orders of magnitude, in sequencing throughput with a corresponding drop in cost per base. A simple exercise to comprehend the scale of acceleration in sequencing is to look back at the state of the art of sequencing in 1980. At that time, earlier improvements in Sanger and Maxam-Gilbert methodologies had initiated the wide use of sequencing in research laboratories around the world. Then, typical sequencing throughputs per slab gel run were under 10,000 bp. During the period from 1980 to 2005 sequencing platforms based on Sanger chemistry had a 500 to 1,000-fold increase, to more than 5 Mbp per run. The number of reads that could be processed, quality, read length and analysis all improved and were optimized, propelled by the development of the human genome project (Barnhart, 1989). While these technological advances were certainly impressive, they dwarf when compared to the acceleration in sequencing capacity after 2005. At that time, novel ultra-high throughput technologies started to become commercially available. From 2005 through the second half of 2011, the throughput per run had increased an additional 100,000-fold, or 5 orders of magnitude. This acceleration has been unprecedented in science and technology. It has outpaced Moore's law that famously predicted that the number of transistors in a computer processor would double every two years (Moore, 1965, Figure 1). This fast increase in sequencing capacity has had important consequences in analysis and logistics, and has changed expectations in all aspects of plant genetics, breeding and biotechnology.

The new chemistries and platforms, broadly described as Next-Generation sequencing (NGS) technologies, take advantage of diverse chemistries and detection approaches. While some of these technologies appear to have little in common with each other, they share key characteristics. All NGS technologies are massively parallel systems relying on the immobilization of millions of, up to billions of DNA templates in a solid surface. They do not use electrophoresis, relying instead on *in situ* base detection and extension. With the exception of one system, developed by Helicos, NGS platforms need to amplify the templates and use one of several PCR-based approaches. One additional characteristic of NGS systems that took more than one early-adopting institution unprepared, is the increased need for computer power and storage necessary to process and retain the massive data produced. Currently there are 5 companies commercializing one or more NGS platforms. However, there are only three NGS technologies, Roche 454, Illumina and ABI SOLiD, that account for the vast majority of usage in plant research and are widely available in academic institutions, private research centers and service-providing companies. As it will be emphasized in the next sections, these platforms have different input and output characteristics that make them more or less advantageous to specific applications. Finally, one 'Third Generation' sequencing platform has recently become commercially available from Pacific Biosciences. Third generation technologies are also massively parallel systems although they use single-molecule DNA templates, real-time detection and are able to generate longer reads faster. The expectation is that third generation machines will eventually produce large numbers of high-quality reads with an average of several kilobase-pairs from a single molecule.

This section provides a brief description and comparison of the most relevant sequencing technologies available in plant biotechnology and breeding. For additional, in-depth technical information, the reader is advised to refer to also other reviews available, including those from Glenn (2011), and Metzker (2010).

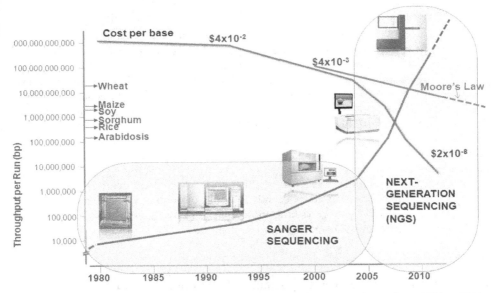

Fig. 1. Increase in maximum throughput per run in sequencing platforms from 1980 to 2011. Throughput per run based on Stratton *et al* (2009) and Glenn (2011).

2.1 Sanger sequence analyzers

For more than 30 years and until recently, sequencing based on the Sanger and Maxam-Gilbert chemistries were the only practical methods to routinely determine DNA sequences in plants and other biological systems. During the 80's and 90's, Sanger-based platforms increased throughput by orders of magnitude, and became the method of choice, while the Maxam-Gilbert method remained a low-throughput process. The development of automated Sanger systems was greatly facilitated by technical innovations such as thermal cycle-sequencing and single-tube reactions in combination with fluorescence-tagged terminator chemistry (Trainor, 1990). Additional improvements in parallelization, quality, read length, and cost-effectiveness were achieved by the development of automatic base-calling and capillary electrophoresis. In the current version of Sanger sequencing a mixture of primer, DNA polymerase, deoxinucleotides (dNTPs) and a proportion of dideoxynucleotide terminators (ddNTP), each labeled with a different fluorescent dye, are combined with the DNA template. During the thermal cycling reaction, DNA molecules are extended from templates and randomly terminated by the occasional incorporation of a labeled ddNTP. DNA is then cleaned up and denatured. Detection is achieved by laser excitation of the fluorescent labels after capillary-based electrophoresis separation of the extension products. The differences in dye excitation generate a "four color" system that is easily translated by a computer to generate the sequence. Modern Sanger sequencers like the Applied Biosystems ABI3730 have reached a high level of sophistication and can achieve routine read-lengths close to 900 bp and per-base 'raw' accuracies of 99.99% or higher (Shendure & Ji, 2008). The ABI 3730xl analyzer can run 96 or 384 samples every 2-10 hours, generating approximately 100,000 bases of raw sequence at a cost of a few hundred dollars.

2.2 Roche 454

The 454 platform (now owned by Roche) was the first NGS platform available as a standalone system. DNA templates need to be prepared by emulsion PCR and bound to beads, with 1–2 million beads deposited into wells in a titanium-covered plate. The Roche 454 technology is based on Pyrosequencing and additional beads that have sulphurylase and luciferase attached to them are also loaded into the same wells to generate the light production reaction. DNA polymerase reactions are performed in cycles but, unlike Sanger, there are no terminators. Instead, one single dNTP is alternated in every cycle in limiting amounts. Fluorescence after the reaction indicates the incorporation of the specific dNTP used in the cycle (Metzker, 2010). Because the intensity of the light peaks is proportional to the number of bases of the same type together in the template, the fluorescence can be used to determine the length of homopolymers, although accuracy decreases considerably with homopolymer length. The current 454 chemistry is able to produce the longest reads of any NGS system, about 700 bp, approaching those generated by Sanger reads. However, 454 systems can sequence several megabases for less than 100 dollars.

2.3 Illumina

The Solexa platform (now owned by Illumina) has become the most widely used NGS system in Plant biotechnology and breeding. Illumina captures template DNA that has been ligated to specific adapters in a flow cell, a glass enclosure similar in size to a microscope slide, with a dense lawn of primers. The template is then amplified into clusters of identical molecules, or polonies, and sequenced in cycles using DNA polymerase. Terminator dNTPs in the reaction are labeled with different fluorescent labels and detection is by optical fluorescence. As only terminators are used, only one base can be incorporated in one cluster in every cycle. After the reaction is imaged in four different fluorescence levels, the dye and terminator group is cleaved off and another round of dye-labeled terminators is added. The total number of cycles determine the length of the read and is currently up to 101 or 151, for a total of 101 or 151 bases incorporated, respectively. At the time of writing this review, this technology was able to yield the highest throughput of any system, with one of the highest raw accuracies. One major disadvantage is the short read it produces. However, paired-end protocols virtually double the read per template and facilitate some applications that were originally out of the reach of the technology. The Illumina HiSeq 2000 sequencer is currently able to sequence up to 540-600 Gbp in a single 2-flow cell, 8.5-day run at a cost of about 2 cents per Mbp (http://www.illumina.com/systems/hiseq_2000.ilmn).

2.4 Life Technologies SOLiD

ABI (now part of Life Technologies) has commercialized the SOLiD (Support Oligonucleotide Ligation Detection) platform. This platform is based on Sequencing by Ligation (SbL) chemistry. SbL is a cyclic method but differs fundamentally from other cyclic NGS chemistries in its use of DNA ligase instead of polymerase, and two-base-encoded probes instead of individual bases as units. In SbL, a fluorescently labeled 2-base probe hybridizes to its complementary sequence adjacent to the primed template and ligated. Non-ligated probes are then washed away, followed by fluorescent detection. In SOLiD, every cycle (probe hybridization, ligation, detection, and probe cleavage) is repeated ten times to yield ten color calls spaced in five-base intervals. The extension product is removed and additional ligation rounds are performed with an n–1 primer, which moves the calls by one position. Color calls from the ligation rounds are then ordered into a linear sequence to decode the DNA sequence (Metzker, 2010). SOLiD has similar throughput and cost per base to Illumina. It also has the best raw accuracy among commercial NGS systems.

2.5 Life Technologies Ion Torrent

Ion torrent is the commercial name for a new NGS platform now owned by Life Technologies (Rothberg et al, 2011; http://www.iontorrent.com). At the time of writing this chapter, the system was not widely used in plant research and its use elsewhere had been described in a limited number of published research papers (e.g. Miller et al, 2011). However, with recent upgrades, fast turnaround times and affordability, the system is finding its way into research laboratories. Currently, its usefulness in being evaluated for a number of applications in plant biotechnology and breeding. Ion Torrent differs from other NGS in that its chemistry does not require fluorescence or chemiluminescence, and for that matter optics (e.g. a CCD camera) to work. Beads, each carrying PCR clones from a single original fragment, are subjected to polymerase synthesis using standard dNTPs on an ion chip. The ion chip is a massively parallel semiconductor-sensing device that contains ion-sensitive, field-effect transistor-based sensors (tiny pH meters, essentially), coupled to more than one million wells where the polymerization reaction occurs. Cycles of reactions including one single nucleotide are produced, in a way that is analogous to the Roche 454 system. In each cycle, the electronic detection of changes in pH due to the release of a proton during base incorporation indicates that a base has been incorporated. The IonTorrent has the lowest throughput but also the fastest turnaround times of all commercially available NGS systems. The current Ion Torrent chip can yield several hundred thousand reads with an average length of about 100 bp in less than 2 hours.

Platform	3730xl	5500xl SOLiD	454 FLX Titanium	HiSeq 2000	GAIIx	MiSeq	Ion Torrent
Company	ABI	ABI	Roche	Illumina	Illumina	Illumina	Life Tech.
Chemistry	Dideoxy	SbL	PS	SbS	SbS	SbS	pH
Amplification	Biol/PCR	EmPCR	EmPCR	BrPCR	BrPCR	BrPCR	EmPCR
Detection	Fluor.	Fluor.	Fluor.	Fluor.	Fluor.	Fluor.	pH
Run Time (days)	0.08	8	0.5	8	14	1.1	0.08
Max. Aver. Length (bp)	900	60x2	700	101x2	151x2	151x2	100
Max. TP/run (Gbp)	0.00008	310	0.8	600	100	1	0.1
Max.Reads/Run(Million)	0.000096	5,167	1	3,000	320	3	1
TP per 24hr (Gbp)	0.00064	45	1	75	7	1	2.4
Raw Error range (%)	0.01	0.01	1-3	0.1	0.1	0.1	(1)*
Equip.Cost (xUS$1,000)	150	600	300	690	350	125	60
Cost per Mbp (US$)	4,000	0.05	8	0.02	0.1	0.7	10

SBS: Sequencing by synthesis, SbL: Sequencing by ligation, PS: Pyrosequencing, EmPCR: Emulsion PCR, Biol: Biological cloning, Fluoresc.: Fluorescence, BrPCR: Bridge PCR, TP: Throughput.

Table 1. Comparison of current sequence technologies.

2.6 Other NGS platforms, Helicos Heliscope, Polonator

There are other NGS systems that have been marketed in the last few years, however, they have had limited use in plant sciences. Helicos developed the first commercial single-molecule sequencer, called HeliScope. However, very few units were sold due to the cost of the machine, on-site requirements and other considerations. Currently, Helicos provides sequencing as a service. One additional company, Azco-Biotech is marketing the Max-Seq Genome sequencer (http://www.azcobiotech.com/instruments/maxseq.php). This commercial version of the academic, open-source Polonator can run either sequencing by synthesis or sequencing by ligation protocols, similar to Illumina and SOLiD, respectively, although it generates shorter reads, 35- or 55-bp-long.

2.7 Pacific Biosciences and the 3rd generation

Pacific Biosciences has launched the PacBio RS platform, considered the first commercially available 3rd-Generation system. The first early-access instruments were deployed in late 2010 and the first commercial batch became available by mid-2011. The PacBio system is based on SMRT, a single-molecule sequencing chemistry with real time detection. The sequencing cell has DNA polymerases attached to nanowells and exposed to single molecule templates and labeled NTPs. No terminators are used, although conditions are set to slow polymerization to a level that can be detected by a CCD camera. Each dNTP has a unique fluorescent label that is detected and then cleaved off during synthesis. Polymerization is detected as it happens, several bases per second. Because of this real-time detection and the enzyme processivity, this method has the potential to generate reads in excess of 10 kilobases in a few minutes. The potential of a technology that is able to sequence single molecules and produce long reads is immense. However, the PacBio technology may need to overcome a number of technical challenges before it reaches a widespread use in plant sciences. Average read length in current outputs exceeds 1 Kbp although single-pass error rate has been reported to be 15%, considerably higher than other sequencing platforms (Glenn, 2011). One major source of errors consists of deletions produced during detection. As will be discussed later, improvements in raw quality and further gains in read length will broaden the range of optimal applications for PacBio.

3. Applications in plant research

Sequencing platforms have different combinations of throughput, cost, read length, number of reads and raw accuracy. Their effective use in plant research and development programs depends on matching the best Sanger, NGS or Third Generation platform to specific applications (Morozova & Marra, 2008; Schuster, 2008; Varshney et al, 2009). One common misconception about Sanger-based systems is that they have, or will soon become obsolete. On the contrary, Sanger capillary systems are still the most widely used sequencers in routine molecular biology applications and are not likely to disappear in the near future. While their number of optimal applications has decreased, Sanger sequencers remain essential in many. The characteristics of capillary Sanger systems make them better suited for confirmatory sequencing in recombinant DNA technology, when the need to determine specific targets at low throughput makes them cost-effective. They are also best in low- to medium-throughput low-complexity shotgun and targeted sequencing experiments, where the use of highly-parallel random sequencing is impractical. Currently, no other chemistry or technology can match Sanger's combination of length and quality that remain the gold standard of sequencing.

Most sequencing applications can be divided into 2 categories: 1)*de novo* sequencing, and 2) resequencing. In the case of *de novo* sequencing, reads are obtained from an unknown sequence and either assembled to reconstruct this sequence or compared directly to reads from other unknown sequences. In the case of resequencing, reads are mapped or aligned to a known reference sequence. *De novo* applications are usually slower and more computer-intensive than resequencing, but are needed to reconstruct genomes and transcriptomes in species with unknown genomes. Major resequencing applications include polymorphism discovery and transcription profiling. This section emphasizes the use of new massive sequencing technologies and how they have recently been deployed in *de novo* and resequencing applications in plant research.

3.1 Physical maps and reference genomes

It is not surprising that considerable effort has been given during the last 15 years to the sequencing of plant genomes. The determination of nuclear and organellar genomes enables the identification of genes, regulatory elements, and the analysis of genome structure. This information improves our understanding of the role of genes in development and evolution, and facilitates the discovery of related genes and functions across species (Messing & Llaca, 1998; Feuillet *et al*, 2011). Reference genomes are also important tools in the identification, analysis and exploitation of genetic diversity of an organism in plant population genetics and breeding (Varshney *et al*, 2009; Edwards & Batley, 2010; Jackson *et al*, 2011). The sequencing of the human genome and other vertebrates in the 90's provided the technological pathway for the initial sequencing of genomes in plants (International Human Genome Sequencing Consortium, 2001, Venter *et al*, 2001). However, the structure of plant genomes poses additional challenges. Plant genomes are characterized by higher proportions of highly repetitive DNA and by the presence of segmental duplications or full genome duplications due to polyploidization events. The 1C genome content in Maize, for example, is smaller than in humans but consists of higher proportions and larger tracks of high-copy elements such as retrotransposable elements. Only a small fraction of the genome corresponds to exons and regulatory regions, usually in low-copy DNA islands that harbor single genes or small groups of genes (Schnable *et al*, 2009; Llaca *et al*, 2011). The average genome size in plants is larger than humans, approximately 5.8 Gbp, and they have a wider size distribution than mammals (Bennett & Leitch, http://data.kew.org/cvalues.). Some important crops like hexaploid wheat can have genomes that are more than 4 times the size of the human genome.

The first completed reference plant genomes, Arabidopsis and Rice, were from model plant species with small genomes, approximately 4% and 12% the size of the human genome. The genomes were produced by Sanger-based shotgun sequencing of overlapping bacterial artificial chromosomes (BACs) (The Arabidopsis Genome Initiative, 2001; International Rice Genome Project, 2005). This BAC-by-BAC approach requires the initial construction, fingerprinting and physical mapping of large numbers of random BACs (Soderlund *et al*, 1997; Ding *et al*, 2001). A subset of BACs is selected based on a minimum tiling path and shotgun libraries are individually constructed from each BAC and completed by subclone end-sequencing and assembly. Finally, BAC sequences are completed using a targeted approach aimed at closing sequencing gaps and finishing low-quality regions. This process, albeit slow and time consuming produced the only two references considered finished to date. These projects were performed by large collaborative consortia and took several years of fingerprinting and sequencing work. The cost of the Arabidopsis genome project has been estimated at US$70 million (Feuillet *et al*, 2011). In maize, a draft reference genome was completed from the inbred line B73 using a similar approach, although no gap closure or low quality finishing steps were completed. The maize draft genome, a highly valuable genetic resource available to the plant research community, was accomplished by multiple laboratories at an estimated cost of tens of millions in a joint NSF/DOE/USDA program. The three BAC-by-BAC sequencing projects mentioned above benefited from working in small units (BACs), which minimized problems associated by misassembly of highly repetitive DNA. One important consideration about BAC-by-BAC genomes is that they are not really complete. They have representation gaps in regions that are "unclonable" under the conditions used to prepare the BAC libraries. Many of these unclonable regions correspond to tandem repeats such as telomeric sequences and other repetitive regions, although it may also include gene space (Schnable *et al*, 2009). Furthermore, even in BAC-by-BAC approaches, the complexity of many plant genomes of moderate size such as maize prevent the creation of a complete physical assembly and there are some regions that may still lay in unassigned regions.

The high cost, long time, and logistics of BAC-by-BAC projects led many groups to adopt an alternative strategy also previously implemented in humans and other vertebrates: Whole-Genome Sequencing (WGS; Venter *et al*, 2001). In WGS, whole genomic DNA is randomly sheared and the fragments are end-sequenced and assembled. This strategy has improved with the use of multiple genomic libraries with different insert sizes and improved assembly software, which can identify such constraints in clone size. Not surprisingly, the first WGS, Sanger-based draft genomes were obtained from small genomes with relatively small amounts of repetitive DNA, including *Populus* (Tuskan *et al*, 2006), Grape (Jailon *et al*, 2007), and Papaya (Ming *et al*, 2008). More recent refinements enabled the sequencing of larger genomes such as *Sorghum bicolor* (~730 Mbp; Paterson *et al*, 2009) and soybean, an ancestral tetraploid (1.1 Mbp; Schmutz *et al*, 2010). The cost and time to accomplish these projects is reduced in comparison to BAC-by-BAC projects, although still considerable. In the case of soybean, the largest plant genome completed by Sanger WGS, sequencing was done by a team of 18 institutions and a total of more than 15 million Sanger reads were produced and assembled from multiple libraries with average sizes ranging from 3.3 Kb to 135 kb (Schmutz *et al*, 2010). In general, WGS approaches are effective in the determination of gene space in small and medium size plant genomes. However, reduction in time and cost is achieved at the expense of assembly fidelity in repetitive regions and expanded need for computational resources. WGS-based approaches increase potential assembly artifacts due to haplotype and homeolog collapse in regions with high identity. This may lead to large numbers of scaffolds to be mapped.

The use of NGS platforms in WGS projects has improved the ability to rapidly determine reference genomes at the expense of overall assembly quality, especially in high copy and duplicated regions. The potato reference genome (The Potato Genome Sequencing Consortium, 2011) was successfully constructed using a combination of Illumina, 454 and

Sanger reads. The implementation of hybrid methods using Roche 454 sequencing in combination with Sanger sequences has been effective in reducing overall cost and time to generate high-quality sequences in gene space regions (Rounsley, 2009). Examples of hybrid references are cucumber (Huang *et al,* 2009) and apple (Velasco *et al, 2010*). The use of NGS-only WGS assemblies, especially based on Illumina or Solid reads, can reduce cost and time by orders of magnitude in relation to Sanger or Hybrid strategies. Medium-size genomes such as maize, can be covered up to 200-fold in a single 9-day run in an Illumina HiSeq2000 system for under $30,000, for example. However, correct mapping and *de novo* assembly of these shotgun short reads has been problematic. Short reads have raised concern about their ability to accurately assemble genomes with high abundance of near identical repetitive sequences and gene duplication. The difficulty of using shotgun short read data for *de novo* assembly has also been a challenge in humans and other animals, but it is exacerbated in plants due to the higher proportion of highly repetitive DNA, segmental duplications and polyploidization. However, improvements have been made recently by using strategies that rely on paired-end reads and mate pairs, the use of multiple libraries with different insert sizes and the development of software with algorithms use end-sequence distance information from these libraries. Using these strategies, contig size, particularly in gene-rich regions, has increased considerably (Li *et al*, 2010). As read length in NGS continues to expand (e.g. Illumina platforms can perform 150-bp paired ends, and Roche 454 has released a long read chemistry), assembly will be improved.

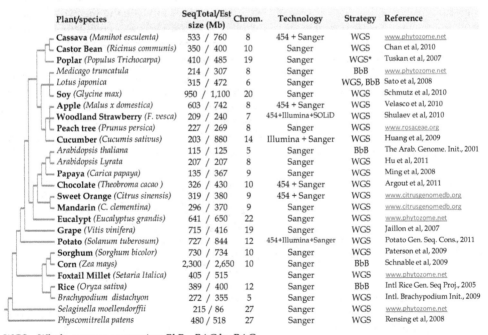

Plant/species	SeqTotal/Est size (Mb)	Chrom.	Technology	Strategy	Reference
Cassava (*Manihot esculenta*)	533 / 760	8	454 + Sanger	WGS	www.phytozome.net
Castor Bean (*Ricinus communis*)	350 / 400	10	Sanger	WGS	Chan et al, 2010
Poplar (*Populus Trichocarpa*)	410 / 485	19	Sanger	WGS*	Tuskan et al, 2007
Medicago truncatula	214 / 307	8	Sanger	BbB	www.phytozome.net
Lotus japonica	315 / 472	6	Sanger	WGS, BbB	Sato et al, 2008
Soy (*Glycine max*)	950 / 1,100	20	Sanger	WGS	Schmutz et al, 2010
Apple (*Malus x domestica*)	603 / 742	8	454 + Sanger	WGS	Velasco et al, 2010
Woodland Strawberry (*F. vesca*)	209 / 240	7	454+Illumina+SOLiD	WGS	Shulaev et al, 2010
Peach tree (*Prunus persica*)	227 / 269	8	Sanger	WGS	www.rosaceae.org
Cucumber (*Cucumis sativus*)	203 / 880	14	Illumina + Sanger	WGS	Huang et al, 2009
Arabidopsis thaliana	115 / 125	5	Sanger	BbB	The Arab. Genome. Init., 2001
Arabidopsis Lyrata	207 / 207	8	Sanger	WGS	Hu et al, 2011
Papaya (*Carica papaya*)	135 / 367	9	Sanger	WGS	Ming et al, 2008
Chocolate (*Theobroma cacao*)	326 / 430	10	454 + Sanger	WGS	Argout et al, 2011
Sweet Orange (*Citrus sinensis*)	319 / 380	9	454 + Sanger	WGS	www.citrusgenomedb.org
Mandarin (*C. clementina*)	296 / 370	9	Sanger	WGS	www.citrusgenomedb.org
Eucalypt (*Eucalyptus grandis*)	641 / 650	22	Sanger	WGS	www.phytozome.net
Grape (*Vitis vinifera*)	715 / 416	19	Sanger	WGS	Jaillon et al, 2007
Potato (*Solanum tuberosum*)	727 / 844	12	454+Illumina+Sanger	WGS	Potato Gen. Seq. Cons., 2011
Sorghum (*Sorghum bicolor*)	730 / 734	10	Sanger	WGS	Paterson et al, 2009
Corn (*Zea mays*)	2,300 / 2,650	10	Sanger	BbB	Schnable et al, 2009
Foxtail Millet (*Setaria Italica*)	405 / 515		Sanger	WGS	www.phytozome.net
Rice (*Oryza sativa*)	389 / 400	12	Sanger	BbB	Intl Rice Gen. Seq Proj., 2005
Brachypodium distachyon	272 / 355	5	Sanger	WGS	Intl. Brachypodium Init., 2009
Selaginella moellendorffii	215 / 86	27	Sanger	WGS	www.phytozome.net
Physcomitrella patens	480 / 518	27	Sanger	WGS	Rensing et al, 2008

WGS – Whole genome sequencing, BbB – BAC by BAC.

Fig. 2. Reference Plant Genomes. The list includes genomes publicly available.

The use of NGS to sequence genomes in a BAC-by-BAC, or a pooled BAC approach can be facilitated by the use of new physical mapping technologies such as whole genome profiling (WGP). This process allows the physical mapping of BACs using a restriction-based fingerprinting approach analogous to high information content gel electrophoresis. In this system BAC clones are pooled, then DNA from the pools are prepared and digested with a restriction endonuclease. Tags are then added to the ends of the fragments and the labeled fragments are end-sequenced using Illumina chemistry. The sequence data are processed and analyzed by an optimized FPC software program to build BAC contigs across the genome. The sequence data obtained during WGP can be combined with BAC, BAC pool, and or Whole Genome Sequencing data. (Steuernagel 2009; van Oeveren *et al*, 2011).

Regardless the assembly strategy or sequence technology used, the completion of reference genomes for most plants remains a big challenge. All publicly available completed references indicated in Figure 2 correspond to plant species with below-average genome sizes. The challenge of sequencing full genomes with vast amounts of duplications and continuous high copy transposable elements remains inaccessible with the current technology. The major technological breakthrough required here is the improvement of 3rd Generation technologies able to produce long reads. Such long reads can then be used to improve contig length in combination with other technologies, or by themselves.

3.2 Development of pan-genomes

The significant sequence diversity and the high structural polymorphism observed in important plant models such as maize, highlights a serious limitation in the use of a single reference genome as a sufficient representative of a species. There is accumulating evidence that large differences in 1C DNA content observed between closely-related species, or between subspecies, landraces and lines in the same species correspond not only to differences in repetitive, non-coding DNA but also to gene content (Morgante *et al*, 2007; Llaca *et al*, 2011). Deep resequencing and the addition of *de novo* assembly of non-reference genomes is necessary to capture gene space included in larger structural variations (i.e. CNVs, PAVs, and large indels). With improvements in long-read sequencing technologies and assembly software and strategies, the creation of reference "pan-genomes" for certain species will be an important resource in plant genetics research

3.3 Genome surveys and partial genome assembly

The use of genome survey sequence (GSS) and partial targeted *de novo* assembly strategies can be useful in gene discovery research projects involving non-model plants species, or species with significant sequence diversity and high structural polymorphism. Maize and other crops exhibit pan-genomes that can be considerably larger than the standard available reference genome, with presence-absence variation (PAV) polymorphisms including both non-genic regions and gene space. Partial *de novo* assembly is also useful for gene discovery in large genomes or any non-model species where a region of interest can be genetically mapped and a partial physical map can be derived from the genetic map. Target regions may be included in one single BAC clone or in a series of overlapping BAC clones, determined by fingerprinting or by known probes that are used to develop assemblies. Rounsley *et al*. (2009) sequenced and assembled a 19Mbp region of chromosome 3 in rice, using the 454 reads and was able to generate scaffolds with size ranging from 243 Kbp to 518 Kbp. Similar approaches have been used in cacao (Feltus *et al,* 2011) and Barley (Steuernagel *et al,* 2009).

3.4 Plant-associate genomics

Genome sequence information from plant pathogen, comensal and mutualistic species is an important resource for plant improvement. Knowledge on gene content, expression and diversity of plant-associated organisms helps our understanding of the basis of their interactions with plants and in developing strategies to modify such interactions. Sequencing the genomes of mutualistic endophytes can help in the modification of nitrogen fixation and other processes, which are essential for developing a more sustainable agriculture. Genomic resources created from pathogens and their non-pathogenic relatives provide not only targets to develop increased disease resistance and pest control, but improved mechanisms for gene transfer in plant biotechnology as well (Wood *et al,* 2001).

Due to their simplicity, DNA and RNA viruses were among the first pathogens to be sequenced. Strains for more than 700 plant viruses, ranging from 1.2 to 30 Kb, have been completely sequenced to date (http://www.ncbi.nlm.nih.gov/genomes/). However, most recent advances in plant-associated genomes have been related to bacterial pathogens. They represent an important data mining resource for plant pathologists regarding some of the most devastating agricultural diseases (Stavrinides, 2009). From a technical point of view, they are also simpler to sequence than plant genomes and better suited for current technologies. Most bacterial genomes are approximately 5 Mbp, approximately 1,000-fold smaller than an average plant genome, with a relatively simple structure, often consisting of a single amplicon, although other amplicons such as megaplasmids and plasmids are frequently present. Bacterial genomes have little or no highly-repetitive elements. Due to their importance and relative technical simplicity, a considerable number of bacterial pathogens had already been sequenced by Sanger methodologies before the advent of NGS. Among the first published genomes are the citrus chlorosis agent *Xylella fastidiosa* (Simpson *et al,* 2000), *Pseudomonas syringae* pv tomato DC3000 (Buell *et al, 2003*), and *Ralstonia solanacearum* (Salanoubat *et al,* 2002). Currently, there are approximately 50 public finished and draft genomes from pathogenic bacteria, plus an even larger number currently in progress or unpublished. The smallest genome is that of *Phytoplasma asteris*, an obligated intracellular pathogen (0.9 Mbp; Oshima *et al,* 2004) and the largest (10.4 Mbp) is from *Streptomyces scabies* 87-22, the causal agent of Potato scab (http://www.sanger.ac.uk/). Complete genomes of other, non-pathogenic, mutualistic bacteria have also been sequenced. Krause *et al* (2006) reported the complete 4.3 Mbp genome of *Azoarcus* sp. strain BH72. Azoarcus is a mutualistic endophyte in cereal species, supplying biologically fixed nitrogen to its host while colonizing plants in high numbers without eliciting disease. Unlike related species that are pathogens, Azoarcus shows a lack of genes that pathogenic bacteria possess that degrade plant cell walls.

Fungi and stramenopiles are eukaryotic groups that include important plant pathogens and mutualists and have been the focus of genome sequencing projects. Stramanemopiles such as *Phytophtora spp.* are fungus-like eukaryotes although they are more closely related to diatoms. Genomes from at least 15 fungal and stramenopile pathogens are publicly available (see the Phytopathogen Genomics Resource, http://cpgr.plantbiology.msu.edu/ for a comprehensive list). Among them is *Phytophtora infestans*, the stramanemopile causal agent of the Irish Potato Famine in the 19th century. The 240 Mbp sequence was assembled using Sanger sequencing (Haas *et al*, 2009). Martin *et al* (2008) sequenced the 65-Mbp genome of the fungus *Laccaria bicolor* that is part of a mycorrhizal symbiosis. The adoption of NGS for *de novo* sequencing of prokaryotic and fungal plant pathogens has been effective, especially when using a combination of 454 and Illumina reads (Reindhart *et al, 2009*; DiGuistini *et al, 2009*).

The biological relevance and economic importance of certain nematodes and insects has made them desirable targets for genome sequencing. Currently, publicly available genomes have been produced by Sanger-based WGS, including drafts for the Northern root-knot nematode *Meloidogyne hapla*, (Opperman *et al*, 2008). The genomes of the Aphid *Acyrthosiphon pisum* and the beetle *Tribolium castaneum*, which produce damage in crops and stored grains, have been completed (International Aphid Genomics Consortium, 2010; Tribolium Genome Sequencing Consortium, 2008).

Within two degrees of separation, genome sequencing of bacterial and fungal species that are pathogens of plant-infesting insects and nematodes are important resources in developing effective and safer strategies for pest resistance. One of such resources is the complete sequencing of replicons comprising the genome of *Bacillus thuringiensis*, which expresses insecticidal crystal proteins that have been used to engineer insect-resistant crops (Roh *et al*, 2007; Challacombe *et al*, 2007).

3.5 Metagenomics

The previous section underscores the importance of understanding the genomics of plant-associated microbiota. However, there are multiple interactions between plants and non-characterized microorganisms, some of them still to be discovered, that cannot be grown in cultures in the laboratory. (Riesenfeld *et al, 2004*; Allen *et al, 2009*). There is increasing evidence of the effect these organisms have in traits such as disease resistance and nitrogen utilization (Handelsman, 2004). Metagenomic studies of these microbial communities can exploit the availability of DNA amplification techniques and highly-parallel, clone-free NGS sequencers to sequence part of their genomes (Chen & Pachter 2005; Leveau, 2007). There are two major roles that high-throughput sequencing technologies can play in metagenomes applied to agriculture. The most common role is the mass sequencing of environmental (e.g. soil, water) samples to provide a systems-biology view of the microbiota under study. This type of study focuses on the genetic diversity and interactions between large numbers of plant associates and plants (Krober *et al, 2009*). Roche 454 Pyrosequencing of small subunit (16S) ribosomal RNA amplicons (pyrotags) is a method for profiling microbial communities that provides deep coverage with low cost, although it is complicated by several artifacts, including chimeric sequences caused by PCR amplification and sequencing errors. Illumina protocols have also been developed for the sequencing of "itags" derived from 16S hypervariable regions, for deep metagenomics analysis (Degnan & Ochman, 2011).

A second trend in modern metagenomics involves its exploitation for the discovery of biomolecules with novel properties. Current discovery strategies involve the screening of metagenomic libraries. Jin *et al* (2007) identified a novel EPSP gene with high resistance to glyphosate and potential use in plant biotechnology by screening a metagenomic library derived from a glyphosate-polluted area. However, the use of ultra-high throughput sequencing that could lead to simpler strategies is currently limited by the length of reads in NGS systems. Long reads are needed to generate full-sequence information within a single read. The use of sequencing approaches in biomolecule discovery will be feasible with the improvement of longer-read, 3rd-Generation technologies such as PacBio.

3.6 Genomic variant discovery for marker development

Linkage mapping, diversity and evolutionary studies in plants rely on the ability to identify and analyze single nucleotide and insertion-deletion polymorphisms (SNPs and Indels), which can be directly related to differences in a phenotype of interest, be genetically linked to its causative factor, or indicate relationships between individuals in populations (Rafalski, 2002). The implementation of high-throughput PCR-based marker technologies (e.g. Taqman) and improvements in Sanger sequencing throughput increased the limits for both the number of markers as well as samples in marker-related studies. These changes have enabled new applications in linkage and association mapping analysis, marker assisted selection (MAS) and characterization of germplasm. They have also facilitated fingerprinting and determination of seed purity. More recently, the emergence of NGS has enabled genome-wide discovery of polymorphisms on a massive scale. The Roche 454 system has been used effectively for polymorphism discovery (Gore *et al*, 2009a), although the higher throughput and lower cost of Illumina and SOLiD technologies make them particularly well suited for major programs when a reference genome is available (Deschamps & Campbell, 2010).

In species such as Arabidopsis and Rice, which have a small genome and an available reference genome, NGS-based genome-wide variant discovery can be simply accomplished by WGS (Ossowski *et al*, 2008; Huang *et al*, 2010). In medium- to large-sized genomes, where the proportion of gene space is reduced and much of the sequence is repetitive, the use of reduced-representation strategies can improve cost effectiveness. Reduced representation strategies involve the selection of specific regions of the genome to reduce complexity and increase coverage for the selected regions. Several enrichment strategies can be used to reduce genome representation. These approaches can utilize previous knowledge about the genome or region of interest. Examples of knowledge-driven enrichment include multiplex long-range PCR, molecular inversion probes (MIP), and sequence capture (Mamanova *et al*, 2010). These methods are usually preferred when a specific region or gene family is targeted. However, random approaches based on restriction digestion and transcriptome sequencing are more adequate in most genome-wide projects (Deschamps & Campbell, 2010). The use of methylation-sensitive enzymes or endonucleases that preferentially cut in low copy DNA have been particularly successful when used in strategies to identify large sets of SNPs in maize and soybean varieties (Gore *et al*, 2009b; Deschamps *et al*, 2009; Hyten *et al*, 2010). Illumina-based SNP discovery strategies using reduced representation libraries (RRLs) have been described by Deschamps and colleagues in soybean. By combining a 6-bp methylation-sensitive and one 4bp-restriction endonuclease they demonstrated enrichment for gene space, and considerable reduction of repetitive DNA reads (Deschamps *et al*, 2009; Fellers, 2008). Other methyl-filtration methods for reduced representation consist of digesting DNA with the endonuclease mcrBC (Gore *et al*, 2009; Palmer *et al*, 2003). Sequences derived from cDNA have been used in both Sanger and NGS sequences to reduce representation for polymorphism detection (Barbazuk *et al*, 2007; Trick *et al*, 2009). One major advantage to this method is the direct targeting of exonic DNA, which increases the chance of detecting functional SNPs, especially when used in conjunction with cDNA normalization methods. However, it can also constrain SNPs within a relatively small number of genes expressed in the tissue and stage used. Both standard whole-genome and RRL approaches usually yield a massive amount of polymorphisms, at a scale that is beyond prior Sanger-based projects. For example, Lam *et al* (2010) reported the genome resequencing in 31 wild and cultivated soy varieties, which led to the identification of more than 10 million SNPs in total, where more than 1 million of them were in genic regions. Nelson *et al* (2011) resequenced 8 sorghum (*Sorghum bicolor*) accessions using a reduced-representation approach in an Illumina system and identified 283,000 SNPs. With seemingly unlimited numbers of SNPs, current bottlenecks have been shifted from the discovery phase to marker assay development and validation.

In plant species where high-quality reference genomes are not available, variant discovery using an NGS resequencing approach can still be accomplished by using alternative references, such as high-quality transcriptome assemblies (see section 3.9) or *de novo* partial assemblies of individual BACs or BAC contigs (see section 3.5). However, both strategies carry some additional risk and validation must include potential for detection of repetitive sequences of paralogous genes. An alternative is an annotation-based strategy, as described in the wheat relative *Aegilops tauschii* by You *et al*. (2011). The genome size is more than 4Gbp and has a large proportion (more than 90%) of repetitive DNA. They produced Roche 454 shotgun reads at low genome coverage from one genotype and identified single-copy sequences and repeat junctions from repetitive sequences as well as sequences shared by paralogous genes. SOLiD and Solexa reads were then generated from another genotype and were mapped to the annotated Roche 454 reads. In this case, 454 reads provide a DNA "context" surrounding the putative SNPs, which can be used to generate genome-wide markers. They were able to identify nearly 500,000 SNPs with a validation rate higher than 81%.

3.7 QTL and eQTL mapping, hapmaps and WGAS

In plants, most agronomically important traits are quantitative. Plant yield, flowering time, sugar content, disease resistance and fruit weight are examples of quantitative traits, which result from the segregation of many genes and are influenced by environmental interactions (Paran & Zamir, 2003). While quantitative traits have been studied for more than 100 years, the mapping of the underlying quantitative trait loci (QTL) could only be accomplished after the development of sequencing methodologies, molecular markers and improved statistical methods. Furthermore, until 2005, only a small fraction of mapped plant QTL had been cloned (Salvi and Tuberosa, 2005; Frary, 2000). One major difficulty was the low resolution of available mapping strategies. Before the advent of NGS platforms, most QTL identification and cloning was based on linkage mapping strategies. In linkage mapping, polymorphisms are identified between two parents and then followed in a large segregating population. The linkage of different regions of the genome to the individual phenotypes can be then inferred statistically by identifying recombinants that show phenotypic differences in the trait of interest. One drawback to linkage mapping is the low resolution that results from relatively few recombinants generated from two original parents in a limited number of generations. Even in cases of QTL with large effects in the total genetic variance, intervals can encompass a large genetic and physical distance and require walking through several megabase-pairs of sequence, with a large number of potential candidates (Yano et al. 1997; El-Assal et al. 2001; Liu et al. 2002). In maize, recent linkage mapping studies have identified QTL with relatively large effects in oil content (Zheng et al, 2008) and root architecture (Ruta et al, 2010).

The development of high-throughput genotyping technologies and later the emergence of NGS platforms has enabled the use of genome-wide association studies (GWAS) and bulked segregant analysis to map plant QTL (Rafalski, 2010; Schneeberger & Weigel, 2011). Unlike linkage mapping, GWAS exploit the natural diversity generated by multi-generational recombination events in a population or panel (Risch & Merikangas, 1996; Yu & Buckler, 2006; Belo et al, 2008; Nordborg & Weigel, 2008). These results in increased resolution compared to linkage mapping populations, as long as enough markers are provided: GWAS may require hundreds of thousands or even millions of genetic markers to achieve sufficient coverage. Before NGS, such marker density was unfeasible and linkage disequilibrium, or association, mapping studies needed to focus on polymorphisms in candidate genes that were suspected to have roles in controlling phenotypic variation for one specific trait of interest (Thornsberry et al, 2001). In plants, availability of NGS and the ability to create lines of individuals with identical or near identical background offer the potential to create public GWAS resources that can be accessed by multiple groups and rapidly resolve complex traits. Plant GWAS can be performed in large numbers of samples in replicated trials using inbreds and recombinant inbred lines (RILs) (Zhu et al, 2008). One or more research groups can then analyze one or many traits in multiple environments. The most important GWAS resource in maize is a collection of recombinant inbred lines derived from a nested association mapping (NAM) population (Gore et al, 2009). The maize NAM population is a collection of 5,000 RILs in sets of 200, derived from one of 25 populations. (Each of the 200 RILs is derived from one F2 plant from a cross between one of 25 inbred lines to B73.) The original inbred lines that were used as founders of the NAM have been resequenced using a NGS reduced-representation approach. Such resequencing surveys (HapMaps) include a high quality data set consisting of 1.4 million SNPs and 200,000 indels spanning the 5,000 inbred lines. Seeds from the RILs can be used to grow and phenotype plants for any trait of interest (McMullen et al, 2009). Recent studies, all derived from the same NAM resource, demonstrate the effectiveness of this approach to identify and characterize QTL. Buckler et al (2009) identified 50 loci that contribute to variation in the genetic architecture of flowering time, with many loci showing small additive effects. Tian et al (2011) also identified large numbers of QTL with small effects determining leaf architecture. Poland et al (2011) identified candidate genes for resistance to northern leaf blight in 29 loci, which included QTL with small additive effects. Kump et al (2011), identified QTL for Southern Leaf blight. HapMaps in Rice have been reported by Huang et al (2010) by resequencing at low coverage a total of 517 landraces that yielded a total of 3.6 million SNPs. The study identified QTL with minor and major contributions to phenotypic variance for drought tolerance, spikelet number and 12 additional agronomic traits. In Medicago truncatula, Branca et al (2011) detected more than 3 million SNPs in 26 inbred lines to study the genetics of traits related to symbiosis and nodulation. In Arabidopsis thaliana, the 1,001 Genomes Project, started in 2008 aims at discovering polymorphisms in that number of wild accessions (Weigel and Mott, 2009; http://1001genomes.org/). The complete genome sequences of over 80 accessions have already been released and inbred lines have been generated from each accession.

Finally, the determination of genome variants and transcription profiling by NGS approaches can be used effectively in the determination of expression quantitative loci (eQTLs; Damerval *et al*, 1994). Variation in the expression of transcripts, when measured across a segregating population, can be used to map regions with *cis*- and *trans*- effects (Holloway & Li, 2010). The development of massively parallel sequencing technologies has replaced microarrays as the method of choice for eQTL analsysis (Holloway *et al*, 2011; West *et al*, 2007). Using NGS, Swanson-Wagner *et al* (2009) identified ~4,000 eQTLs in reciprocal crosses between the maize inbred lines B73 and Mo17, most of them acting in *trans* and regulated exclusively by the paternally transmitted allele.

3.8 Genotyping by sequencing

The value of NGS-driven massive polymorphisms discovery can be seriously restricted by cost and time limitations in the design, validation and deployment of molecular markers. With the falling cost of NGS there is an increased interest in genotyping-by-sequencing (GbS), where the obtained sequence differences are used directly as markers for analysis. As described in section 3.7, Maize NAM and other GWAS comunity-based resources already make use of GbS. However, such panels have limited utility beyond their populations or panels. A number of reduced-representation GbS protocols have been reported that can be applied to other population or panels for linkage, association, bulked segregant analysis, fingerprinting, diversity and other studies. Depending on the details of the project and the available resources, sequences can be mapped to a reference. However, in large genomes or other genomes with no reference available, the consensus of reads flanking the polymorphism can be used as a partial reference or polymorphic reads can simply be treated as dominant markers (Elshire *et al*, 2011). Construction of a low-density GbS linkage map using Restriction-Site-Associated DNA (RAD) has been reported in barley (Chutimanitsakun *et al*, 2010). The use of simpler and highly multiplexed protocols, however, is required in most cases to make GbS cost and time-effective. The bottom line is that an all-inclusive cost per sample is lower than those provided by other available genotyping platforms. Cost estimates need to include the considerable computational and bioinformatic resources needed for GbS data analysis. Using a simple reduced-representation procedure based on *Ape*KI restriction digestion, Elshire *et al* (2011) identified and mapped approximately 200,000 polymorphisms in the 2 parents and the 276 RILs from the maize IBM (B73 x Mo17) mapping population at an estimated cost of $29.00 per sample. With the same protocols and using a single Illumina run, they can process up to 672 samples, taking the actual data collection cost to well under US$20.00 per sample. The low cost per base and the high numbers of reads produced per run make the Illumina and SOLiD systems more suitable for GbS.

3.9 Transcriptome assembly and profiling

The sequencing of DNA products synthesized from total and mRNA isolates (cDNA) has been crucial in gene expression analysis, discovery and determination of alternative splicing forms of genes (isoforms). In the case of organisms with a genome sequence available, cDNA sequencing has facilitated the annotation of splicing sites and untranslated regions (UTRs), as well as improved gene prediction algorithms (Brautigam & Gowik, 2010). As indicated before, transcriptome sequencing can also be deployed as a reduced-representation strategy to identify polymorphisms for marker development and genotyping. Before the advent of NGS, multiple Sanger sequencing strategies were developed for the quantitative and qualitative analysis of mRNA expression. The need for direct quantitative analysis on gene expression led to the development of profiling strategies such as Serial-Analysis-of-Gene-Expression (SAGE; Velculescu *et al*, 1995). On the other hand, the creation of large consortia dedicated to providing end-sequence for individual clones from cDNA libraries enabled gene discovery, annotation and expression on a large scale (Rafalski *et al*, 1998). These efforts have yielded more than 22 million Sanger-based expressed sequenced tags (ESTs) from more than 40 plant species. The largest datasets correspond to arabidopsis, soybean, rice, maize and wheat, each of which contains more than 1 million entries (http://www.ncbi.nlm.nih.gov/dbEST/). There are also EST databases available for at least 42 fungal, stramenopile and nematode phytopathogen species (http://cpgr.plantbiology. msu.edu). Sequence information derived from these databases has been utilized to develop expression microarrays, which are aimed at establishing relative abundance of known genes in large numbers of samples and tissues within the same species or among related species (Rensink and Buell, 2005). While such arrays have been effective in providing gene expression data, they are inheritably biased in their design and have limitations in resolution and in their ability to differentiate between individual genes within families. The highly parallel, short-read NGS technologies such as Illumina and SOLiD have allowed the development of transcription profiling strategies that are more sensitive and accurate than SAGE or microarrays. Initial NGS strategies for transcription profiling had their roots in the innovative, now obsolete massively parallel signature sequencing (MPSS) technology. This technology, owned and provided as a service by Lynx Therapeutics, consisted in the generation and sequencing of short 17-bp unique tags, or signatures, from 3'-UTRs of transcripts at high coverage (Simon *et al*, 2009). It provided unparalleled resolution generating over a million signature sequences per experiment, although the cost of every experiment was considerable (Reinartz *et al*, 2002). With Illumina, tag-based "digital" expression profiling protocols became relatively simple and achieved higher resolution than MPSS at a fraction of the cost (Wang *et al*, 2010) .

The use of shotgun sequencing of cDNA using the Roche 454 analyzer has provided relatively long reads and high coverage for gene discovery, annotation and polymorphism discovery in both model and non-model plant species (Barbazuck *et al*, 2007; Emrich *et al*, 2007). More recently, the increasing gains in throughput, as well as improvement in shotgun RNA sequencing (RNA-seq) strategies and analysis software have expanded the potential of Illumina and SOLiD platforms for full transcriptome analysis, and replaced the use of the tag-based expression profiling approach. In RNA-seq, total or messenger RNA is fragmented and converted into cDNA. Alternatively, it is first converted into cDNA and then fragmented. Adaptors are attached to one or both ends, and sequenced as single- or paired-ends (Wang *et al*, 2009; Margerat & Bahler, 2010). Depending on the genomic resources available for the organism of interest, the resulting sequences can be aligned to either a reference genome or reference transcripts. Alternatively, genes can be assembled *de novo*. In either case, cDNA sequencing provides considerably more information on the transcriptome, including gene structure, expression levels, presence of multiple isoforms and sequence polymorphism. Unlike microarray-based hybridization, it does not depend upon previous knowledge of potential genes. A considerable number of RNA-Seq projects have been made in major crop species. Severin *et al* (2010), Zenoni *et al* (2010) and Zhang *et al* (2010), all applied Illumina-based RNA-Seq on multiple tissues and stages in soy, grape and rice, respectively, and aligned transcript reads to their respective reference genome sequences. Li *et al* (2011) used Illumina in multiple stages along a leaf developmental gradient and in mature bundle sheath and mesophyll cells.

3.10 Small RNA characterization

Small RNAs (sRNA) are non-protein-coding small RNA molecules ranging from 20 to 30 nt that have a role in development, genome maintenance and plant responses to environmental stresses (Simon *et al, 2009*). Most sRNAs belong to two major groups: 1) microRNAs (miRNA) are about 21 nt and usually have a post-transcriptional regulatory role by directing cleavage of a specific transcript, 2) short interfering RNAs (siRNA) are usually 24 nt-long and influence *de novo* methylation or other modifications to silence genes (Vaucheret 2006). The finding of their prevalence in low-molecular-weight fractions of total RNA in animals and plants predated the development of NGS. However, the use of MPSS greatly expanded resolution and later became clear that short-read NGS technologies such as Illumina or SOLiD had optimal characteristics in sRNA analysis (Zhang *et al*, 2009). Roche 454 sequencing has also been used in sRNA analysis (Gonzales-Ibeas *et al*, 2011).

3.11 Epigenomics

In plants and other multicellular organisms, cell differentiation is driven by variation of epigenetic marks encoded on the DNA or chromatin. Such variations can be stable, or heritable, but do not change the underlying DNA sequence of the genome (Zhang & Jeltsch, 2010). Furthermore, there is accumulating evidence that transgenerationally inherited epigenetic variants (epialleles) have a significant effect in differential gene expression in plant and animal populations (Reinders *et al*, 2009; Johannes *et al*, 2009). Biochemical alterations such as cytosine methylation polymorphisms and differential histone deacetylation are epigenetic marks that can play a critical role in development (Schöb & Grossniklaus, 2006; Chen & Tian, 2007). In plants, 5-methylcytosine can be present at symmetric CG sites but also can be located at CHG sites as well as in asymmetric CHH locations (where H can be A, C or T). Methylation at CG sites usually occurs symmetrically on both strands and is heritable, maintained by specific types of methyltransferases that recognize hemimethylated sites created during replication. Methylation of CHG and CHH is established and maintained by additional methyltransferases (Schob & Grossniklaus, 2006). One important consideration is that both the epigenome and methylome will be considerably larger than the genome of an organism. As a major part of the epigenome, the methylome consists of the sum of genome and methylation states at every cytosine location. Multiple states coexist in the same individual, depending on cell types, tissues, developmental stages or environments. Adding additional complexity is the fact that methylation at one position may be partial within the same cell type. Similar to the transcriptome, methylome analysis has, therefore, a quantitative component in addition to a qualitative one.

Different Sanger and NGS strategies have been developed over the years that can directly or indirectly identify epigenetic marks and patterns. Before NGS, epigenetic studies were mostly limited to individual genes or sets of candidate genes or regions. One exception is the extensive work done in arabidopsis by Zhang *et al* (2007), which provided the first genome-wide study in plants and considerable information on methylation distribution and effect in gene expression. The use of NGS technologies coupled with bisulfite conversion, restriction digestion, or immunoprecipitation strategies are facilitating genome-wide methylome analysis in plants. Of these, approaches based on sodium bisulfite conversion (BC) provide the highest resolution. In BC, denatured DNA is treated with sodium bisulfite, which induces the hydrolytic deamination of cytosine. Subsequent treatment with a desulfonation agent transforms the uracyl-sulfonate intermediate into uracyl. Replication and further amplification of the converted strands will incorporate a thymidine in originally non-methylated C sites. However, 5-methylcytosine residues remain unreactive during conversion and further amplification will retain the original CG pairing at that position (Liu *et al*, 2004; Frommer *et al* 1992). As a consequence, unmethylated and methylated cytosines can be mapped in the two original strands and resolution can reach single base level, as long as the original sequence is known (Zhang *et al*, 2007; Lister *et al*, 2008). The combination of methylome BC-NGS and high-definition transcriptome analysis will have an important role in further characterization of epi-regulation in plants.

4. Future outlook

For more than 30 years, the use of sequencing methods and technologies, in combination with strategies in breeding and molecular genetic modification, has contributed both to our knowledge of plant genetics and to remarkable increases in agricultural productivity. In recent years, agricultural sciences have been in the middle of a second technological revolution in DNA sequencing, driven in large part by the post-human genome goal of affordable genome sequencing for disease research and personalized medicine. The resulting NGS systems have become a "disruptive technology", radically reducing limitations in sequence information and consequently altering the types of questions and problems that can be addressed (Mardis, 2010). As we have explored in this chapter, these massively parallel sequencing systems have had a dramatic effect in variant and gene discovery, genotyping, and in the characterization of transcriptomes, genomes, epigenomes and metagenomes. The increasing ability to sequence complete genomes from multiple individuals within the same species is providing a more comprehensive view of crop diversity and transgenesis, as well as a better understanding of the effect of mutational and epimutational processes in plant breeding. With all their high throughput and low cost, different NGS platforms have specific flaws and researchers have been playing on their weaknesses and strengths. Short reads, even at very high throughputs, may not be the best output for *de novo* sequencing. (After all, longer Sanger reads are not ideal either to determine the most complex eukaryotic genomes.) Alternatively, the use of systems that are able to produce longer reads but have relatively lower capacity will probably not be effective in some of the most important future applications in molecular breeding such as genotyping-by-sequencing.

The boundaries of sequencing technologies continue to expand, and the quest for more universal sequencers continues. The expected improvements in quality and read length in real-time 3rd-Generation systems have the potential to greatly benefit plant *de novo* sequencing and metagenomics applications. The development of cheaper, portable, easier-to-use machines has the potential to create a decentralization of sequencing and to create entirely new field applications. With shorter technological cycles in sequencing and computing, it is difficult to anticipate the next disruptive technology. While it is likely that 3rd Generation systems will soon become widespread in plant research, there has also been progress toward nanopore-based technologies. Nanopore systems are based on electronic detection of DNA sequence and have the potential of low sample preparation work, high speed, and low cost (Branton *et al*, 2008). Future improvements in sequencing technologies will enable applications that can support discovery and innovation needed to respond to growing population pressure, energy crises, decreasing fresh water availability and climate change (Gepts & Hancock, 2006; Moose & Mumm, 2008). Ultimately, DNA sequencing is one of a number of tools in plant breeding and biotechnology, albeit an essential one. The knowledge of genomic organization, diversity and function of genes in crops needs to be associated to an understanding of plant biology. Effective plant breeding programs need solid statistical strategies and the ability to create, manage and integrate large, heterogeneous sets of data based on phenotype and sequence-derived information.

5. Acknowledgments

The author thanks Antoni Rafalski, Stephane Deschamps, Delia Ines Lugo and Karlene Butler for their critical reading of this manuscript and for many interesting discussions.

6. References

Allen, C., Bent, A. & Charkowski, A. (2009). Underexplored niches in research on plant pathogenic bacteria. *Plant Physiology* 150(4): 1631-1637.

Arabidopsis Genome Initiative (2001). Analysis of the genome sequence of the flowering plant *Arabidopsis thaliana*. *Nature* 408(6814): 796-815.

Argout, X. *et al* (2011). The genome of *Theobroma cacao*. *Nature Genetics* 43(2): 101-108.

Barbazuk, W.B. *et al* (2007). SNP discovery via 454 transcriptome sequencing. *Plant Journal*, Vol.51(5): 910–918.

Barnhart, B.J. (1989). DOE Human Genome Program. *Human Genome Quarterly* 1(1): 1.

Bedbrook, J.R., Smith, S.M. & Ellis, R.J. (1980). Molecular cloning and sequencing of cDNA encoding the precursor to the small subunit of chloroplast ribulose 1,5-bisphosphate carboxylase. *Nature* 287(5784): 692–697.

Belo, A. *et al* (2008). Whole genome scan detects an allelic variant of fad2 associated with increased oleic acid levels in maize. *Molecular Genetics & Genomics* 279(1): 1-10.

Bernatsky, R. & Tanksley, S. (1986). Toward a saturated linkage map in tomato based on isozymes and random cDNA sequences. *Genetics* 112(4): 887-898.

Branca, A. *et al* (2011). Whole-genome nucleotide diversity, recombination, and linkage disequilibrium in the model legume *Medicago truncatula*. *Proc. Natl. Acad. Sci. USA* 108(42): E864–E870.

Branton, D. *et al* (2008). The potential and challenges of nanopore sequencing. *Nature Biotechnology* 26(10): 1146-1153.

Brautigam, A. & Gowik, U. (2010). Next generation sequencing as a valuable tool in plant research. *Plant Biology* 12(6): 831-841.

Buckler, E.S. *et al* (2009). The architecture of maize flowering time. *Science* 325(5941): 714-718.

Buell, C.R. *et al* (2003). The complete genome sequence of the Arabidopsis and tomato pathogen *Pseudomonas syringae* pv. tomato DC3000. *Proc. Natl. Acad. Sci. USA* 100(18): 10181–10186.

Challacombe, J.F. *et al* (2007). The complete genome sequence of *Bacillus thuringiensis* Al Hakam. *Journal of Bacteriology* 189(9): 3680-3861.

Chan, A.P. *et al* (2010). Draft genome sequence of the oilseed species *Ricinus communis*. *Nature Biotechnology* 28(9): 951-956.

Chen, K. & Pachter, L. (2005). Bioinformatics for whole-genome shotgun sequencing of microbial communities. *PLoS Computational Biology* 1(2): 106-112.

Chen, Z.J., Tian, L. (2007). Roles of dynamic and reversible histone acetylation in plant development and polyploidy. *Biochimica et Biophysica Acta* 1769(5-6): 295-307.

Chutimanitsakun, Y. *et al* (2011). Construction and application for QTL analysis of a Restriction Site Associated DNA (RAD) linkage map in barley. *BMC Genomics* 12(4): 1-13.

Damerval, C. *et al* (1994). Quantitative trait loci underlying gene product variation: A novel perspective for analyzing genome expression. *Genetics* 137(1): 289-301.

Degnan, P.H. & Ochman, H. (2011). Illumina-based analysis of microbial community diversity. *The ISME Journal*. doi:10.1038/ismej.2011.74. [Epub ahead of print].

Deschamps, S. *et al* (2010). Rapid genome-wide single nucleotide polymorphism discovery in soybean and rice via deep resequencing of reduced representation libraries with the Illumina Genome Analyzer. *The Plant Genome* 3(1): 53-68.

Deschamps, S. & Campbell, M. (2010). Utilization of next-generation sequencing platforms in plant genomics and genetic variant discovery. *Molecular Breeding* 25(4): 553–570.

Diguistini, S. *et al* (2009). *De novo* genome sequence assembly of a filamentous fungus using Sanger, 454 and Illumina sequence data. *Genome Biology* 10(R94): doi:10.1186/gb-2009-10-9-r94.

Ding, Y. *et al* (2001). Five-color-based high-information-content fingerprinting of bacterial artificial chromosome clones using type IIS restriction endonucleases. *Genomics* 74(2): 142-154.

Edwards, D. & Batley, J. (2010). Plant genome sequencing: applications for crop improvement. *Plant Biotechnology Journal* 8(1): 2-9.

El-Assal, S.E.D. *et al* (2001). The cloning of a flowering time QTL reveals a novel allele of CRY2. *Nature Genetics* 29(4): 435–440.

Elshire, R.J. *et al* (2011). A robust, simple genotyping-by-sequencing (GBS) approach for high diversity species. *PLoS One* 6(5): e19379.

Emrich, S.J. *et al* (2007). Gene discovery and annotation using LCM-454 transcriptome sequencing. *Genome Research* 17(1): 69-73.

Fellers, J.P. (2008). Genome filtering using methylation-sensitive restriction enzymes with six-base pair recognition sites. *The Plant Genome* 1(2): 146-152.

Feltus, F.A. *et al* (2011). Sequencing of a QTL-rich region of the *Theobroma cacao* genome using pooled BACs and the identification of trait specific candidate genes. *BMC Genomics* 12(379): 1-16.

Feuillet, C. *et al* (2011). Crop genome sequencing: lessons and rationales. *Trends in Plant Science* 16(2): 77-88.

Fraley, R.T. *et al* (1983). Expression of bacterial genes in plant cells. *Proc. Natl. Acad. Sci. USA* (80)15: 4803–4807.

Frary, A. *et al* (2000). fw2.2: A quantitative trait locus key to the evolution of tomato fruit size. *Science* 289(5476): 85-88.

Frommer, M. *et al* (1992). A genomic sequencing protocol that yields a positive display of 5-methylcytosine residues in individual DNA strands. *Proc. Natl. Acad. Sci. USA* 89(5): 1827–1831.

Gepts, P. & Hancock, J. (2006). The future of plant breeding. *Crop Science* 46(4): 1630–1634.

Glenn, T.C. (2011). Field guide to next-generation DNA sequencers. *Molecular Ecology Resources* 11(5): 759-769.

Gonzalez-Ibeas, D. *et al* (2011). Analysis of the melon (*Cucumis melo*) small RNAome by high-throughput pyrosequencing. *Plant Physiology* 150(4): 1631–1637.

Gore, M.A. *et al* (2009a). A First generation haplotype map of maize. *Science* 326(5956): 1115-1117.

Gore, M.A. *et al* (2009b). Large-scale discovery of gene-enriched SNPs. *The Plant Genome* 2(2): 121-133.

Haas, B.J. *et al* (2009). Genome sequence and analysis of the Irish potato famine pathogen *Phytophthora infestans*. *Nature* 461(7262): 393-398.

Handelsman, J. (2004). Metagenomics: Application of genomics to uncultured microorganisms. *Microbiology and Molecular Biology Reviews* 68(4): 669-685.

Herrera-Estrella, L. *et al* (1983). Expression of chimaeric genes transferred in to plant cells using a Ti plasmid-derived vector. *Nature* 303: 209-213.

Holloway, B. & Li, B. (2010). Expression QTLs: applications for crop improvement. *Molecular Breeding* 26(3): 381-391.

Holloway, B. *et al* (2011). Genome-wide expression quantitative trait loci (eQTL) analysis in maize. *BMC Genomics* 12(336): 1-14.

Hu, T.T. *et al* (2011). The *Arabidopsis lyrata* genome sequence and the basis of rapid genome size change. *Nature Genetics* 43(5): 476-481.

Huang, S. *et al* (2009). The genome of the cucumber, *Cucumis sativus* L. *Nature Genetics* 41(12): 1275-1281.

Huang, X. *et al* (2010). Genome-wide association studies of 14 agronomic traits in rice landraces. *Nature Genetics,* 42(11): 961-967.

Hyten, D.L. *et al* (2010). High-throughput SNP discovery through deep resequencing of a reduced representation library to anchor and orient scaffolds in the soybean whole genome sequence. *BMC Genomics* 11(38): 1-8.

International Aphid Genomics Consortium (2010). Genome sequence of the pea aphid *Acyrthosiphon pisum*. *PLoS Biology* 8(2): e1000313.

International Human Genome Sequencing Consortium (2001). Initial sequencing and analysis of the human genome. *Nature* 409(6822): 860–921.

International Rice Genome Sequencing Project (2005). The map-based sequence of the rice genome. *Nature* 436(7052): 793-800

Jackson, S.A. *et al (2011).* Sequencing crop genomes: approaches and applications. *The New phytologist* 191(4): 915-925.

Jaillon, O. *et al* (2007). The grapevine genome sequence suggests ancestral hexaploidization in major angiosperm phyla. *Nature* 449(7161): 463-467.

Jin, D. *et al* (2007). Identification of a new gene encoding EPSPS with high glyphosate resistance from a metagenomic library. *Current Microbiology* 55(4): 350–355.

Johannes, F. *et al* (2009). Assessing the impact of transgenerational epigenetic variation on complex traits. *PLoS Genetics* 5(6): e1000530.

Krause, A. *et al* (2006). Complete genome of the mutualistic, N2-fixing grass endophyte *Azoarcus sp.* strain BH72. *Nature Biotechnology* 24(11): 1384-1390.

Krober, M. *et al* (2009). Phylogenetic characterization of a biogas plant microbial community integrating clone library 16S-rDNA sequences and metagenome sequence data obtained by 454-pyrosequencing. *Journal of Biotechnology* 142(1): 38-49.

Kump, K.L. *et al* (2011). Genome-wide association study of quantitative resistance to southern leaf blight in the maize nested association mapping population. *Nature Genetics* 43(2): 163-168.

Lam, H.M. *et al* (2010). Resequencing of 31 wild and cultivated soybean genomes identifies patterns of genetic diversity and selection. *Nature Genetics* 42(12): 1053-1059.

Leveau, J.H. (2007). The Magic and menace of metagenomics. *European Journal of Plant Pathology* 119(3): 279–300.

Li, P. *et al* (2011). The developmental dynamics of the maize leaf transcriptome. *Nature Genetics* 42(12): 1060-1067.

Li, R. *et al* (2010). *De novo* assembly of human genomes with massively parallel short read sequencing. *Genome Research* 20(2): 265-272.

Lister, R. *et al* (2008). Highly integrated single-base resolution maps of the epigenome in Arabidopsis. *Cell* 133(3): 523-536.

Liu, J. *et al* (2002). A new class of regulatory genes underlying the cause of pear-shaped tomato fruit. *Proc. Natl. Acad. Sci. USA* 99(20): 13302-13306.

Liu, L. *et al* (2004). Profiling DNA methylation by bisulfite genomic sequencing. Problems and solutions. *Methods in Molecular Biology* 287(1): 169-179.

Llaca, V., Campbell, M. & Deschamps, S. (2011). Genome diversity in maize. *Journal of Botany*, 2011(104172):1-10.

Mamanova, L. *et al* (*2010*). Target-enrichment strategies for next-generation sequencing. *Nature Methods* 7(2): 111-118.

Mardis, E.R. (2010). A decade's perspective on DNA sequencing technology. *Nature* 470(7333): 198-203.

Marguerat, S. & Bahler, J. (2010). RNA-seq: from technology to biology. *Cellular & Molecular Life Sciences* 67(4): 569–579.

Martin, F. *et al* (2008). The genome of *Laccaria bicolor* provides insights into mycorrhizal symbiosis. *Nature* 452(7182): 88-92.

Maxam, A.M. & Gilbert, W. (1977). A new method for sequencing DNA. *Proc. Natl. Acad. Sci. USA* 74(2): 560–564.

McMullen, M.D. *et al* (2009). Genetic properties of the maize nested association mapping population. *Science* 325(5941): 737-740.

Messing, J. & Llaca, V. (1998). Importance of anchor genomes for any plant genome project. *Proc. Natl. Acad. Sci. USA* 95(5): 2017–2020.

Metzker, M.L. (2010). Sequencing technologies - the next generation. *Nature Reviews* 11(1): 31-46.

Miller, W. *et al* (2011). Genetic diversity and population structure of the endangered marsupial *Sarcophilus harrisii* (Tasmanian devil). *Proc. Natl. Acad. Sci. USA* 108(3): 12348-12353.

Ming, R. *et al* (2008). The draft genome of the transgenic tropical fruit tree papaya (*Carica papaya* Linnaeus). *Nature* 452(7190): 991-996.

Moore, G.E. (1965). Cramming more components onto integrated circuits. *Electronics* 38(8): 114-117.

Moose, S.P. & Mumm, R.H. (2008). Molecular plant breeding as the foundation for 21st century crop improvement. *Plant Physiology* 147(3): 969-977.

Morgante, M., De Paoli, E. & Radovic, S. (2007). Transposable elements and the plant pan-genomes. *Current Opinion in Plant Biology* 10(2): 149–155.

Morozova, O. & Marra, M.A. (2008). Applications of next-generation sequencing technologies in functional genomics. *Genomics* 92(5): 255-264.

Nelson, J.C. *et al* (2011). Single-nucleotide polymorphism discovery by high-throughput sequencing in sorghum. *BMC Genomics* 12(135): 1-14.

Nordborg, M. & Weigel, D. (2008). Next-generation genetics in plants. *Nature* 456(7223): 720-723.

Opperman, C.H. *et al* (2008). Sequence and genetic map of *Meloidogyne hapla*: compact nematode genome for plant parasitism. *Proc. Natl. Acad. Sci. USA* 105(39): 14802-14807.

Oshima, K. *et al* (2004). Reductive evolution suggested from the complete genome sequence of a plant-pathogenic phytoplasma. *Nature Genetics* 36(1): 27-29.

Ossowski, S. *et al* (2008). Sequencing of natural strains of *Arabidopsis thaliana* with short reads. *Genome Research* 18(12): 2024-2033.

Palmer, L.E. *et al* (2003). Maize genome sequencing by methylation filtration. *Science* 302 (5653): 2115-2117.

Paran, I. & Zamir, D. (2003). Quantitative traits in plants: beyond QTL. *Trends In Genetics* 19(6): 303-306.

Paterson, A.H. *et al* (2009). The *Sorghum bicolor* genome and the diversification of grasses. *Nature* 457(7229): 551-556.

Poland, J.A. *et al* (2011). Genome-wide nested association mapping of quantitative resistance to northern leaf blight in maize. *Proc. Natl. Acad. Sci. USA* 108(17): 6893-6898.

Rafalski, A. *et al* (1998). New experimental and computational approaches to the analysis of gene expression. *Acta Biochimica Polonica* 45(4): 929-934.

Rafalski, A. (2002). Applications of single nucleotide polymorphisms in crop genetics. *Current Opinion in Plant Biology* 5(2): 94-100.

Rafalski, A. (2010). Association genetics in crop improvement. *Current Opinion in Plant Biology* 13(2): 174-180.

Reinartz, J. *et al* (2002). Massively parallel signature sequencing (MPSS) as a tool for in-depth quantitative gene expression profiling in all organisms. *Briefings in Functional Genomics & Proteomics* 1(1): 95-104.

Reinders, J. & Paszkowski, J. (2009). Unlocking the Arabidopsis epigenome. *Epigenetics* 4(8): 557-563.

Reinhardt, J.A. *et al* (2009). *De novo* assembly using low-coverage short read sequence data from the pathogen *Pseudomonas syringae* pv. oryzae. *Genome Research* 19(2): 294-305.

Rensing, S.A. *et al* (2008). The *Physcomitrella* genome reveals evolutionary insights into the conquest of land by plants. *Science* 319(5859): 64-69.

Rensink, W.A. & Buell CR. (2005). Microarray expression profiling resources for plant genomics. *Trends in Plant Science* 10(12): 603-609.

Riesenfeld, C., Schloss PD, & Handelsman J. (2004). Metagenomics: genomic analysis of microbial communities. *Annual Review of Genetics* 38: 525–552.

Risch, N. & Merikangas, K. (1996). The future of genetic studies of complex human diseases. *Science* 273(5281): 1516-1517.

Roh, J.Y. *et al* (2007). *Bacillus thuringiensis* as a specific, safe, and effective tool for insect pest control. *Journal of microbiology and biotechnology* 17(4): 547–559.

Rothberg, J.M. *et al* (2011). An integrated semiconductor device enabling non-optical genome sequencing. *Nature* 475(7356): 348-352.

Rounsley, S. *et al* (2009). *De novo* Next Generation Sequencing of Plant Genomes. *Rice* 2(1): 35-43.

Ruta, N. *et al* (2010). QTLs for the elongation of axile and lateral roots of maize in response to low water potential. *Theoretical & Applied Genetics* 120(3): 621-631.

Salanoubat, M. *et al* (2002). Genome sequence of the plant pathogen *Ralstonia solanacearum*. *Nature* 415(6871): 497-502.

Salvi, S. & Tuberosa, R. (2005). To clone or not to clone plant QTLs: present and future challenges. *Trends in Plant Science* 10(6): 297-304.

Sanger, F., Nicklen, S. & Coulson, A.R. (1977). DNA sequencing with chain-terminating inhibitors. *Proc. Natl. Acad. Sci. USA* 74(12): 5463-5467.

Sato, S. *et al* (2008). Genome of the legume, Lotus japonicus. *DNA Research* 15(4): 227-239.

Schmutz, J. *et al* (2010). Genome sequence of the palaeop soybean. *Nature* 463(7278): 178-183.

Schnable, P.S. *et al* (2009). The B73 maize genome: complexity, diversity, and dynamics. *Science* 326(5956): 1112-1115.

Schneeberger, K. & Weigel, D. (2011). Fast-forward genetics enabled by new sequencing technologies. *Trends in Plant Science* 16(5): 282-288.

Schöb, H. & Grossniklaus, U. (2006). The first high-resolution DNA "methylome". *Cell* 126(6): 1025-1028.

Schuster, S.C. (2008). Next-generation sequencing transforms today's biology. *Nature Methods* 5(1): 16-18.

Severin, A.J. *et al* (2010). RNA-Seq Atlas of *Glycine max*: a guide to the soybean transcriptome. *BMC Plant Biology* 10(160): 1-16.

Shendure, J. & Ji, H. (2008). Next-generation DNA sequencing. *Nature Biotechnology* 26(10): 1135-1145.

Shulaev, V. *et al* (2011). The genome of woodland strawberry (*Fragaria vesca*). *Nature Genetics* 43(2): 109-116.

Simon, S.A. *et al* (2009). Short-read sequencing technologies for transcriptional analyses. *Annual Review of Plant Biology* 60: 305-333.

Simpson, A.J. *et al* (2000). The genome sequence of the plant pathogen *Xylella fastidiosa*. *Nature* 406(6792): 151-157.

Soderlund, C., Longden, I. & Mott, R. (1997). FPC: a system for building contigs from restriction fingerprinted clones. *Comparative and Applied Biosciences* 13(5): 523-535.

Stavrinides, J. (2009). Origin and evolution of phytopathogenic bacteria. In: *Plant Pathogenic Bacteria: Genomics and Molecular Biology*. R.W. Jackson (Ed). 330. Caister Academic Press. ISBN 9781904455370. UK.

Steuernagel, B. *et al* (2009). *De novo* 454 sequencing of barcoded BAC pools for comprehensive gene survey and genome analysis in the complex genome of barley. *BMC Genomics* 10(547): 1-15.

Stratton, M.R., Campbell, P.J. & Futreal, P.A. (2009). The Cancer Genome. *Nature* 458(7239): 719-724.

Swanson-Wagner, R.A. *et al* (2009). Paternal dominance of trans-eQTL influences gene expression patterns in maize hybrids. *Science* 326(5956): 1118-1120.

The international Brachypodium Initiative. (2010). Genome sequencing and analysis of the model grass *Brachypodium distachyon*. *Nature* 463(7282): 763-768.

The Potato Genome Sequencing Consortium. (2011). Genome sequence and analysis of the tuber crop potato. *Nature* 475(7355): 189-195.

Thornsberry, J.M. *et al* (2001). Dwarf8 polymorphisms associate with variation in flowering time. *Nature Genetics* 28(3): 286-289.

Tian, F. *et al* (2011). Genome-wide association study of leaf architecture in the maize nested association mapping population. *Nature Genetics* 43(2): 159-162.

Trainor, G.L. (1990). DNA sequencing, Automation and the Human Genome. *Analytical Chemistry* 62(5): 418-426.

Tribolium Genome Sequencing Consortium (2008). The genome of the model beetle and pest *Tribolium castaneum*. *Nature* 452(7190): 949-955.

Trick, M. *et al* (2009). Single nucleotide polymorphism (SNP) discovery in the polyploid *Brassica napus* using Solexa transcriptome sequencing. *Plant Biotechnology Journal* 7(4): 334-346.

Tuskan, G.A. *et al* (2006). The genome of black cottonwood, *Populus trichocarpa* (Torr. & Gray). *Science* 313(5793): 1596-1604.

van Oeveren, J. (2011). Sequence-based physical mapping of complex genomes by whole genome profiling. *Genome Research* 21(4): 618-625.

Varshney, R.K. *et al* (2009). Next-generation sequencing technologies and their implications for crop genetics and breeding. *Trends in Biotechnology* 27(9): 522-530.

Vaucheret, H. (2006). Post-transcriptional small RNA pathways in plants: Mechanisms and regulations. *Genes & Development* 20(7): 759-771.

Velasco, R. *et al* (2010). The genome of the domesticated apple (*Malus x domestica* Borkh.). *Nature Genetics* 42(10): 833-839.

Velculescu, V.E. *et al* (1995). Serial analysis of gene expression. *Science* 270(5235): 484-487.

Venter, J.C. *et al* (2001). Sequence of the human genome. *Science* 291(5507): 1304-1351.

Wang, Z., Gerstein, M. & Snyder, M. (2009). RNA-Seq: a revolutionary tool for transcriptomics. *Nature Reviews Genetics* 10(1): 57-63.

Wang, L., Li, P. & Brutnell, T.P. (2010). Exploring plant transcriptomes using ultra high-throughput sequencing. *Briefings in Functional Genomics* 9(2): 118-128.

Weigel, D. & Mott, R. (2009). The 1001 genomes project for *Arabidopsis thaliana*. *Genome Biology* 10(5): 107.

West, M.A. *et al* (2007). Global eQTL mapping reveals the complex genetic architecture of transcript-level variation in Arabidopsis. *Genetics* 175(3): 1441-1450.

Wood, D.W. *et al* (2001). The genome of the natural genetic engineer *Agrobacterium tumefaciens* C58. *Science* 294(5550): 2317-2323.

Yano, M. *et al* (1997). Identification of quantitative trait loci controlling heading date in rice using a high-density linkage map. *Theoretical and Applied Genetics* 95: 1025-1032.

You, F.M. *et al* (2011). Annotation-based genome-wide SNP discovery in the large and complex *Aegilops tauschii* genome using next-generation sequencing without a reference genome sequence. *BMC Genomics* 12(59): 1-19.

Yu, J. & Buckler, E.S. (2006). Genetic association mapping and genome organization of maize. *Current Opinions in Biotechnology* 17(2): 155-160.

Zenoni, S. *et al* (2010). Characterization of transcriptional complexity during berry development in *Vitis vinifera* using RNA-Seq. *Plant Physiology* 152(4): 1787-1795.

Zhang, L. *et al* (2009). A genome-wide characterization of microRNA genes in maize. *PLoS Genetics* 5(11), e1000716.

Zhang, G. *et al* (2010). Deep RNA sequencing at single base-pair resolution reveals high complexity of the rice transcriptome. *Genome Research* 20(5): 646-654.

Zhang, X. *et al* (2007). Genome-wide high-resolution mapping and functional analysis of DNA methylation in arabidopsis. *Cell* 126(6): 1189-1201.

Zhang, Y. & Jeltsch, A. (2010). The Application of Next Generation Sequencing in DNA Methylation Analysis. *Genes* 1(1): 85-101.

Zheng, P. *et al* (2008). A phenylalanine in DGAT is a key determinant of oil content and composition in maize. *Nature Genetics* 40(3): 367-372.

Zhu, C. *et al* (2008). Status and prospects of association mapping in plants. *The Plant Genome Journal* 1(1): 5-20.

6

Improvement of Farm Animal Breeding by DNA Sequencing

G. Darshan Raj

Department of Animal Husbandry and Veterinary Services, Government of Karnataka, India

1. Introduction

Animal breeds with specific qualities like adaptability to local climatic conditions, farming situations , resistance to diseases and those meet the requirements of local needs and perform under available feeds and fodder are valuable and present requirement for the world. Main farm animals cattle, sheep, goat and pigs were domesticated 9,000–11,000 years ago, whereas the dog was domesticated earlier, ~14,000 years ago, and the chicken ~4,500 years ago. Most domestic animals have a much broader genetic basis. Molecular studies have shown that the two main forms of domestic cattle, European–African (*Bos Taurus taurus*) and Asian (*Bos taurus indicus*), originate from two different subspecies of the wild ancestor (Loftus *et al.*, 1994). There are thirty well recognized cattle breeds in India which are highly adoptive to local conditions and each breed has distinctive characters. The world population of buffalo is about 158M in comparison with around 1.3B cattle, 1B sheep and 500-600M goats. There are two types of domestic water buffalo, the River Buffalo (*Bubalus bubalis*) which are more widely spread globally, and are the predominant type found in the west from India to Europe. The second type, the Swamp Buffalo (*Bubalus carabanesis*), is found more frequently in the east from India to the Philippines. The two major types of buffalo can be distinguished by their different karyotypes, the Swamp Buffalo has 48 chromosomes, whereas the River buffalo has 50. However, the two types crossbreed to produce fertile hybrid progeny and so cannot be considered as truly different species.

In India and Pakistan there are well defined breeds, each with their own individual qualities. In South Asia 18 River buffalo breeds are recognized, which can be placed into 5 major groups designated as the Murrah, Gujarat, Uttar Pradesh, Central Indian and South Indian breeds. The Mediterranean Buffalo is of the River type and has been strongly selected for milk production, particularly in Italy. It has been genetically isolated for a long time and as a result has become genetically distinct from other breeding populations. The high level of milk production of the Mediterranean Buffalo has resulted in it been exported to many countries worldwide.

Sheeps belong to the Bovidae family and a ruminant animal. It has 54 chromosomes and average life span 10-15 years. Sheeps are excellent weed destroyer. Sheep Breeds of India are based on the breed distribution in different regions 18 well recognized breeds were identified in India. There are nearly 102 breeds of goats in the world, of which 20 breeds are in India. Goats are reared for two purposes i.e. meat & milk. Meat and Milk production is the main objective. The domestic goat (*Capra aegagrus hircus*) is a subspecies of goat domesticated from the wild goat of southwest Asia and Eastern Europe. The goat is a member of the Bovidae family and is closely related to the sheep as both are in the goat-antelope subfamily Caprinae. Goats are one of the oldest domesticated species. Goats have been used for their milk, meat, hair, and skins over much of the world. In the twentieth century they also gained in popularity as pets.

Farm animal populations harbor rich collections of mutations with phenotypic effects that have been purposefully enriched by breeding. Farm animals are of particular interest for identifying genes that control growth, energy metabolism, development, appetite, reproduction and behavior, as well as other traits that have been manipulated by breeding. Genome research in farm animals will add to our basic understanding of the genetic control of these traits and the results will be applied in breeding programmes to reduce the incidence of disease and to improve product quality and production efficiency (Leif Andersson, 2001).

2. Evaluation of animal traits

Many of the traits of interest in animal production are quantitative traits. Evaluation of genetic merit of animals is still essentially based on the application of the theory of quantitative genetics. Traditionally, the genetic improvement of livestock breeds has been based on phenotypic selection. The past century was characterized by the development of quantitative theory and methodology towards the accurate selection and prediction of genetic response (Walsh, 2000).This resulted in the selection of a number of economically important genetic traits in cattle, sheep, pigs and poultry. The conceptual basis of this theory is the polygenic model, which assumes that quantitative traits result from the action (and interaction) of a large number of minor genes, each with small effect. The resulting effects are then predicted using powerful statistical methods (animal model), based on pedigree and performance recording of traits from the individual animal and its relatives (Yahyaoui, 2003). Most of morphological markers are sex limited, age dependent, and are significantly influenced by the environment. Biochemical markers show low degree of polymorphism. The various genotypic classes are indistinguishable at the phenotypic level because of the dominance effect of the marker and low genome coverage (Montaldo and Meza-Herrera, 1998).

2.1 Genome mapping

The advances in molecular genetics technology in the past two decades, particularly Nucleic acid-based markers, has had a great impact on gene mapping, allowing identification of the underlying genes that control part of the variability of these multigenic traits. Broadly, two experimental strategies have been developed for this purpose: linkage studies and candidate gene approach (Jeffreys et al., 1985).

2.1.1 Linkage studies

Linkage studies rely on the knowledge of the genetic map and search for quantitative trait loci (QTL) by using pedigree materials and comparing segregation patterns of genetic markers (generally micro satellites) and the trait being analyzed. Markers that tend to co-segregate with the analyzed trait provide approximate chromosomal location of the underlying gene (or genes) involved in part of the trait variability determinism (Yahyaoui, 2003).

2.1.2 Candidate gene approach

The candidate gene approach studies the relationship between the traits and known genes that may be associated with the physiological pathways underlying the trait (Liu et al., 2008). In other words, this approach assumes that a gene involved in the physiology of the trait could harbor a mutation causing variation in the trait. The gene or part of gene, are sequenced in a number of different animals, and any variation found in the DNA sequences, is tested for association with variation in the phenotypic trait (Koopaei and Koshkoiyeh, 2011). This approach has had some success. For example a mutation was discovered in the estrogen receptor locus (ESR) which results increased litter size in pig. There are 2 problems with the candidate gene approach. Firstly, there are usually a large number of candidate genes affecting the trait, so many genes must be sequenced in several animals and many association studies carried out in a large sample of animals. Secondly, the causative mutation may lie in a gene that would not have been regarded prior as an obvious candidate for this particular trait. Candidate gene approach is performed in 5 steps: 1) collection of resource population. 2) Phenotyping of the traits. 3) Selection of gene or functional polymorphism that potentially could affect the traits. 4) Genotyping of the resource population for genes or functional polymorphism. Lastly, one is statistical analysis of phenotypic and genotypic data (Da, 2003). This is an effective way to find the genes associated with the trait. So far a number of genes have been investigated. Candidate gene approach has been ubiquitously applied for gene disease research, genetic association studies, biomarker and drug target selection in many organisms from animals (Tabor et al., 2002). The traditional candidate gene approach is largely limited by its reliance on existing knowledge about the known or presumed biology of the phenotype under investigation, unfortunately the detailed molecular anatomy of most biological traits remain unknown (Zhu and Zhao, 2007).

A candidate gene approach has been already successfully applied to identify several DNA markers associated with production traits in livestock The principle is based on the fact that variability within genes coding for protein products involved in key physiological mechanisms and metabolic pathways directly or indirectly involved in determining an economic trait (e.g. feed efficiency, muscle mass accretion, reproduction efficiency, disease resistance, etc.) might probably explain a fraction of the genetic variability for the production trait itself. The first step is the identification of mutations in candidate genes that can be analyzed in association studies in specific designed experiments. For the roles they play, growth hormone (GH) and myostatin (MSTN) genes can be considered candidate genes for meat production traits and casein genes in milk production traits (Rothschild and Soller, 1997).

2.1.2.1 Candidate gene polymorphism and DNA sequencing technique

A typical mammalian genome including most of the farm animal species, contain s around 3×10^9 base pairs contained in a specific number of chromosomes typical of each species. The number of human genes was initially estimated at around 50,000-1, 00,000. However, based on the recent complete sequencing of the human genome in the 2002 year , this number has been revised to 25000-30000 only. A similar estimate has been reported for most of the farm animal species. These genes prescribe the development of the living species and contribute to the biological biodiversity of our planet. A feature of the genome that has confused geneticists for years is that the coding sequences (genes) only represent some 3to 5 percent on the genome. Whereas the rest is sequences (genes) unknown functions including different types of repetitive DNA, introns, 5′ and 3′ untranslated regions and pseudogenes. The sizes of genes vary from less than 100 bp to over 2000kb (Fields et al., 1994).

Recent developments in DNA technologies have made it possible to uncover a large number of genetic polymorphisms at the DNA sequence level and to use them as markers for evaluation of the genetic basis for observed phenotypic variability. These markers posses unique genetic properties and methodological advantages that make them more usefull and amendable for genetic analysis and other genetic markers. The possible applications of molecular markers in livestock industry are by conventional breeding programme to transgenic breeding technologies. In conventional breeding strategies molecular markers have several short range or immediate applications viz., parentage determination , genetic distance estimation , determination of twin zygosity and free martinism, sexing preimplation embryo and identification of genetic disease carrier and long range applications viz ., Genome mapping and marker assisted selection. In transgenic breeding molecular markers can be used as reference points for identification of the animals carrying the trasgenes. The progress in the genetic markers suggests their potential use for genetic improvement in the livestock species.

The progress in the DNA recombinant technology , gene cloning and High throughput Sequencing technique during last two decades have brought revolutionary changes in the field of basic as well as applied genetics which provide several new approaches for genome analysis with greater genetic resolution . it is now possible to uncover a large number of genetic polymorphisms at DNA sequence level and to use them as markers for evaluation of the genetic basis for observed phenotypic variability. The first demonstration of DNA level polymorphism, known as restriction fragment polymorphism (RFLP) an almost unlimited number of molecular markers have accumulated. Currently, more powerful and less laborious techniques to uncover new types of DNA makers are steadily being introduced. The introduction of polymerase chain reaction (PCR) in conjunction with the constantly increasing DNA sequence data also represents a mile stone in this endeavor.

In the past polymorphisms in a candidate gene were routinely detected by PCR –RFLP , which involves the amplification of the gene and digesting the amplicon with restriction enzymes and then using gel electrophoresis to separate resulting fragments. In PCR-RFLP technique the gene difference among animals are detected by whether or not restriction enzymes cuts or not resulting in different sized fragments. The genetic differences are usually due to an SNP within the restriction enzyme recognition site, although there might be genetic differences due to insertions and deletions (ins /del) in the gene, which will also results in fragment size differences , although there is no variation in the restriction enzyme recognition site. Animals are expected to have two alleles for every gene except for X and Y chromosomes in males, so that one fragment on the electrophoresis gel would indicate homozygous for the PCR-RFLP allele, where as two different sized fragments would suggest that an animal is heterozygous. However animal may be misclassified as homozygote if there is polymorphism in the primer sequence, which prevents the allele from being amplified and therefore not being detected on the electrophoresis gel referred to as null allele. A null allele will often be detected when misparentage routinely found for a marker system. An animal might also be misclassified, if another, non allelic form was amplified with the PCR primers and digestion of the PCR product results in a different sized fragments. A Non allelic form is revealed by sequencing the fragments contained within the electrophoresis bands, which is recommended step when establishing marker system (Jiang and Ott, 2010).

However new technologies have significantly advanced our ability to identify SNP's and then explore multiple candidate genes at one time at a much lower cost/polymorphism than the PCR-RFLP method. The identification SNP's within a gene or genetic region is now relatively easy To do this , genomic DNA of key animals within a population are sequenced using high throughput automatic sequencing and then compared with the other sequences within the population or to sequences which are publically available databases .

This approach is already being employed with regard to bovine leukocyte adhesion deficiency (BLAD; Shuster *et al.*, 1992) and genes with major effects, such as the halothane locus in swine (Rempel *et al.*, 1993) and alpha $_{s1}$-casein in goat (Manfredi *et al.*, 1995).Currently, direct sequencing is one of the high throughput methods for mutation detection, and is the most accurate method to determine the exact nature of a polymorphism. Sanger dideoxy-sequencing can detect any type of unknown polymorphism and its position, when the majority of DNA contains that polymorphism. Fluorescent sequencing can have variable sensitivity and specificity in detecting heterozygotes because of the inconsistency of base-calling of these sites (Yan *et al.*, 2000). Thus, it has only limited utility when the polymorphism is present in a minor fraction of the total DNA (for example in pooled samples of DNA or in solid tumors) due to low sensitivity. DNA sequencing is usually used as a second step to confirm and identify the exact base altered in the target region previously identified as polymorphic by using scanning methods.

2.1.2.1.1 Different methods of DNA sequencing

The revolution in molecular genetics, and particularly genome sequencing, has already provided benefits for animal breeding, but in comparison with what the future holds, our present tools will undoubtedly be seen as primitive. Complete genome sequences are available for a number of species, genome sequences for the chicken, cow, horse, mouse and chimpanzee are either completed or nearing completion, and single nucleotide polymorphism (SNP) libraries for these species are growing rapidly. This information will underpin most of the developments in livestock breeding and breed management during the coming two decades A quite recent source of obtaining affordable sequence data are high throughput next-generation sequencing (NGS) technologies (Illumina GA/Solexa, RocheGS/454, Solid, Helicos). The massively parallel approaches in these technologies allows much faster and more cost-effective sequence 16 determination than the traditional dideoxy chain terminator method described by Sanger in 1977. The throughput for these new sequencing approaches is measured in billions of base pairs per run, thousand times more than the daily output of an automated 96 capillary DNA sequencer using dye-terminator sequencing technology. High throughput is achieved by immobilization of 400 thousand (Roche) to 40 million (Illumina) target DNAs and sequencing these fragments simultaneously. Furthermore, these new technologies allow the direct sequencing of DNA or cDNA without any cloning step and at decreasing cost per sequenced base . Although all of these new technologies produced extremely short reads when they initially entered the market, the performance of (NGS) platforms have increased substantially. As an example: the first nextgeneration sequencer released by 454 (GS20) had an average read-length of 110 bp. A second improved sequencer has since then been introduced, the GS-FLX, which is able to obtain average read lengths of 250 bases and is able to perform mate-paired reads. A more recent release is the 454 FLX Titanium which has an average read length of 400 bp

Technology	Approach	Maxthrougput (bps)	Read length	% Accuracy	Paired end	$/Mbp
Sanger ABI3730xl[1]	Synthesis with dye terminators	1 Mbp/day(12)	800	99.0->99,999	no	1000
454/Roche FLX[2]	Pyrosequencing	100Mb/7.5hr (3.7K)	250	96.0-99.5	2x110	30
Illumina/Sole xa[3]	Sequencing by synthesis	3 Gbp/5days(6.9K)	36	96.2-99.7	2x36	2.10
ABI/SOLiD[4]	sequencing by ligation	3Gbp/5days (6.9K)	35	99.0>99.94	2x25	1.30
Helicos[5]	single-molecule sequencing	7.5 Gbp/4days (22K)	25	93->99.0	no	n/a

Table 1. Accuracy of different DNA sequencing methods: Kerstens, (2010).

Throughput is compared by giving the maximum number of bases per second a platform generates at it's optimal read length. Paired-end sequencing is for some platforms restricted to a read length. Sequencing cost is compared by indicating the price per mega base and only includes the costs of reagents and costs to perform one sequencing run.

1. Applied Biosystems http://www.appliedbyosystems.com,
2. Roche Applied Science http://www.roche-applied-science.com,
3. Illumina, Inc. http://www.illumina.com
4. Applied Biosystems http://www.appliedbyosystems.com
5. Helicosbio http://www.helicosbio.com (Kerstens, 2010)

The impact of massive parallel sequencing or next generation sequencing (NGS) in Biology, and hence in Animal Genetics, is difficult to overstate. It is a revolution comparable with the one that followed Sanger sequencing forty years ago. It is not just a dramatic increase in sequencing speed: it means a change in paradigm that obliges researchers, institutions and funding agencies alike. It is also a tremendous and passionate technological race worth millions. Although it is uncertain which, if any, of the extant technologies will prevail in the future, and whether they may be replaced by new technologies, one thing is certain: next generation sequencing is this generation sequencing. Current NGS technologies provide a throughput which is at least 100 times that of classical Sanger sequencing and the technologies are quickly improving. Compared with standard sequencing, these technologies do not need cloning, i.e. they are basically shotgun sequencing and result in shorter sequences, currently from 50 to 400 bp. Further, it is promised that new, third generation single molecule sequencing will soon be available and should deliver complete genome sequence that is in time and pricing affordable for every research group (Perez-Enciso and Ferretti, 2010)

NGS technologies allow cost effective sequencing of multiple individuals which is beneficial for SNP detection. SNP discovery through parallel pyrosequencing (454) of an individual human genome identified 3.32 million SNPs, with 606,797 of those as novel SNPs (Wheeler *et al* ., 2008) which is comparable with the shotgun-sequenced Venter genome (cost $70 million) that had 3.47 million SNPs, with 647,767 of those being novel (Wheeler *et al* ., 2008) .These authors stated that at least a 20X genome coverage is required to call 99% of heterozygous bases correctly within a single individual, resulting in a sequencing cost of 2 million dollar. For the Illumina method similar results (3.07 million SNPs of which 420 thousand novel SNPs) can be obtained by sequencing an individual genome with 36X coverage for less than half a million dollars (Wang *et al* ., 2008) or ~ 4 million SNPs of which 26% are novel SNPs at 30X coverage for a quarter of a million dollar. (Bentley *et al* ., 2008). More recent developments using single molecule sequencing further reduce the sequencing costs. For example, the discovery of 2.8 million SNPs by whole genome resequencing at 28X coverage using this technology reduces the costs to less than 50 thousand dollars (Pushkarev *et al* ., 2009). These developments indicate that extremely high throughput sequencing machines that produce relatively short reads are favorable for SNP discovery in a whole genome resequencing approach. In addition, simulations suggest that 85% of 35 bp reads can be placed uniquely on the human genome whereas 95% of 35 bp paired end reads with 200 bp insert sizes have unique placement (Li *et al.* ,2008). However SNP discovery in species with limited public genomic resources benefits from longer read lengths. It has been demonstrated that reads produced with pyrosequencing technology can be assembled de novo into reasonably long contigs that subsequently can serve as a genome reference on which short reads of other individuals can be mapped to detect SNPs (Novaes *et al.*, 2008). At present whole genome sequence assembly by alignment of relative short NGS fragments without the availability of a reference genome is tedious, but possible for less complex mammalian sized genomes (Li *et al* ., 2009). Besides the consistent pattern of non-uniform sequence coverage each NGS platform generates (Harismendy *et al.*, 2009), which is substantially lower in AT-rich repetitive sequences, repetitive regions are hard to reconstruct by short-read sequence assembly. Possible NGS SNP detection strategies circumventing the requirement of a sequenced reference genome are (ultra short read) sequencing of more than one genotype and alignment of that data by using: (1) genome or transcriptome sequence data from model species closely related to the species of interest, (2) whole transcriptome or reduced representative genome sequence data for the species of interest, based on Roche/454 sequence technology (Wiedmann *et al* ., 2008).

Farm animals are quite valuable as resources, often notable as models for pathology and physiology studies. The reproductive physiology of farm animals is more similar to humans than that of rodents because farm animals have longer gestation and pre-pubertal periods than mice. Specific farm animal physiology, such as the digestive system of the pig is similar to that of humans. These attributes of farm animals reveal that they are an unparalleled resource for research replicating human physiological function. For decades, breeders have altered the genomes of farm animals by first searching for desired phenotypic traits and then selecting for superior animals to continue their lineage into the next generation. This genomic work has already facilitated a reduction in genetic disorders in farm animals, as many disease carriers are removed from breeding populations by purifying selection. By studying diverse phenotype over time, researchers can now monitor mutations that occur as wild species become domesticated (Fadiel *et al.*, 2005).

2.2 Animal genome projects

According to Wikipedia, Genome projects are scientific endeavours that ultimately aim to determine the complete genome sequence of an organism (be it an animal, a plant, a fungus, a bacterium, anarchaean, a protist or a virus) and to annotate protein-coding genes and other important genome-encoded features. The genome sequence of an organism includes the collective DNA sequences of each chromosome in the organism. The release of the first draft of the chicken genome in March 2004 spawned the current boom in chicken genomic research (Antin and Konieczka, 2005) evolutionary standpoint; investigation of the chicken genome will provide significant information needed to understand the vertebrate genome evolution, since the chicken is between the mammal and fish on the evolutionary tree. Furthermore, the chicken remains significant as a food animal which comprises 41% of the meat produced in the world and serves as a reliable model for the study of diseases and developmental biology (Dequeant and Pourquie, 2005). With this sequenced genome, chicken breeders will have a framework for investigating polymorphisms of informative quantitative traits to continue their directed evolution of these species (Fadiel *et al.*, 2005)

The sequencing of the pig genome generated an invaluable resource for advancements in enzymology, reproduction, endocrinology, nutrition and biochemistry research (Wernersson, *et al.*, 2005 and Rothschild, *et al.*, 2003) Since pigs are evolutionarily distinct to both humans and rodents, but have co-evolved with these species, the diversity of selected phenotypes make the pig a useful model for the study of genetic and environmental interactions with polygenic traits (Blakesley *et al.*, 2004). The sequencing of the pig genome is also instrumental in the improvement of human health. Clinical studies in areas such as infectious disease, organ transplantation, physiology, metabolic disease, pharmacology, obesity and cardiovascular disease have used pig models (Rothschild, 2004). In the near future, the sequencing of the porcine genome will allow gene markers for specific diseases to be identified, assisting breeders in generating pig stocks resistant to infectious diseases (Klymiuk, and Aigner, 2005 and Fadiel *et al* ., 2005)

Cattle is of great interest since it represents a group of eutherian mammals phylogenetically distant from primates (Larkin, *et al.*, 2003 and Kumar, and Hedges, 1998). Working with the cow species, B.taurus, is significant because the cow is such an economically important animal. This form of livestock makes up the beef and milk production industry. The identification of numerous single-nucleotide polymorphisms (SNPs) makes it possible for geneticists to find associations between certain genes and cow traits that will eventually lead to the production of superior-quality beef (Adam, 2002).

Bioinformatics researchers from New Zealand, US, UK and Australia have come together to work on the sheep genome map. The focal point of interest in sheep is based on the quest to maximize sheep meat and cotton wool production (Fadiel *et al.*, 2005).These studies have revealed the existence of mutations that yield phenotypes unique to the sheep, emonstrating that genetic analysis of the sheep can enhance our knowledge of biological pathways in other mammalian species (Cockett *et al.*, 2001 and Mouchel *et al.*,2001).

Animal	Publication Year	Journal
Chicken (Gallus sonneratii)	2004	Nature
Bovine (Bos taurus)	2009	Science
Horse (Equus caballus)	2009	Science
Cat (Felis catus)	2007	Genome research
Dog (Canis familiaris)	2005	Nature
Pig	2009	Nature
Sheep	2010	Animal Genetics

Table 2. Important Species with known reference sequences:

Cow	Bos taurus	Mammal	Draft Assembly (7X)	BCM-HGSC	Completed
Cat	Felis catus	Mammal	Draft Assembly	WUGSC	Completed
Chicken	Gallus gallus	Non mammalian vertebrate	Draft Assembly (6 ,6X)	WUGSC	Completed
Dog	Canis familiaris	Mammal	Draft Assembly	BI/MIT	Completed
Horse	Equus caballas	Mammal	Draft Assembly (7X)	BI/MIT	Completed
Pig	Sus scrofa	Mammal	Draft Assembly (BAC to BAC)	Sanger	Completed
Rabbit	Oryctolagus cuniculus	Mammal	Low coverage (~ 2X)	BI/MIT	Completed
Sheep	Ovis aries	Mammal	Draft Assembly	BCM-HGSC	In process
Goat	Caprine	Mammal	Draft Assembly	BCM-HGSC	Completed
Buffaloe	Bubalus bubalis	Mammal	Draft Assembly (BAC to BAC)		In process

BCM –HGSC - Baylor College of Medicine, Houston, Human Genome sequence center USA Wu
Washington University Genomic sequencing Center USA
BI/MIT Broad institute / Massachusetts Institute of Technology center for Genome research
USA.Sangers institute, UK.

Table 3. Livestock Genome sequenced. Eggen, (2008).

2.3 Practical advantages animal breeding by DNA sequencing technology

Advances in genomic technology in recent years were driven primarily by human DNA sequencing and genotyping projects. The human sequence was completed in 2001, the cattle sequence in 2004. Design of the cattle SNP chip required obtaining equally spaced SNP that are highly polymorphic across many dairy and beef breeds (Van Tassell et al., 2008) Breeders now use thousands of genetic markers to select and improve animals. Previously only phenotypes and pedigrees were used in selection, but performance and parentage information was collected, stored, and evaluated affordably and routinely for many traits and many millions of animals. Genetic markers had limited use during the century after Mendel's principles of genetic inheritance were rediscovered because few major QTL were identified and because marker genotypes were expensive to obtain before 2008. Genomic evaluations implemented in the last two years for dairy cattle have greatly improved reliability of selection, especially for younger animals, by using many markers to trace the inheritance of many QTL with small effects. More genetic markers can increase both reliability and cost of genomic selection. Genotypes for 50,000 markers now cost <US$200 per animal for cattle, pigs, chickens, and sheep. Lower cost chips containing fewer (2,900) markers and higher cost chips with more (777,000) markers are already available for cattle, and additional genotyping tools will become available for cattle and other species VanRaden et al., (2011). Livestock selection has used estimated breeding values (EBV) based on phenotypic data and pedigree records for more than 40 years. More recently, advances in molecular genetic techniques, in particular DNA sequencing, have led to the discovery of regions of the genome that influence traits in livestock. However utilizing both sources of data in genetic evaluation schemes, such as BREEDPLAN, has been a challenge due to the heterogeneity of data sources, the multi-trait nature of the evaluations, and unknown effects of the marker information on all traits in the evaluation. The SmartGene for Beef project identified significant effects of the Catapult Genetics GeneSTAR tenderness markers on meat tenderness as recorded by the objective measure of shear force These results have been used to further develop methods for combining EBVs (i.e. phenotypic and pedigree data) and gene marker information into a single marker-assisted EBV called an EBVM. Flight time is an objective measure of an animal's temperament which has been shown to be heritable and moderately genetically correlated with SF, thus representing a potential genetic indicator trait for meat tenderness (Johnston and Graser 2009).

2.3.1 Cattle

2.3.1.1 Dairy animals

An Example for milk trait, Schild and Geldermann (1996), Koczan et al. (1991), and Bleck et al. (1996) identified mutations in the 5' flanking regions of bovine casein genes. In their experiments nine of these mutations were screened by DNA sequencing to see whether they might change affinity to nuclear proteins– transcription factors affect expression of relevant genes and quantitatively influence composition of milk proteins. Eighty-one Polish Black and White (BW) and 195 Polish Red (PR) cows were screened by DNA sequencing for polymorphism. Both breeds were found to differ significantly in the distribution of various genotypes in the 5'-flanking regions of α_{S1}- and α_{S2}-casein genes. No polymorphism was found in the 5'-flanking region of the β-casein gene. However, preferential associations were found between individual promoter genotypes and protein variants of α_{S1}- and β-caseins. It provided strong evidences for the existence of specific haplotype combinations, including both coding and regulatory sequences at the casein locus in the Polish cattle breeds. These results showed that nucleotide sequence variations in the promoter regions of bovine casein genes might change the affinity of these regions to nuclear proteins – transcription factors – and thus affect the expression of relevant genes that quantitatively influence the composition of milk proteins (Martin et al., 2002).

A recent project by scientists at the USDA-ARS Beltsville Agricultural Research Center, a total of 5,369 Holstein bulls and cows that were born from 1952-1999 were genotyped with the BovinSNP50 BeadChip (VanRaden *et al.*, 2009; Cole *et al.*, 2009). Genotypic and phenotypic data of these bulls were used to estimate the effects of 38,416 SNP markers (after discarding markers with low minor allele frequency and markers that were in complete linkage disequilibrium with adjacent markers) on production, type, longevity, udder health, and calving ability. Next, the estimated SNP effects were used to compute the genomic PTA of each of 2,035 young Holstein bulls born from 2000-2003 that had no progeny. Finally, the 2009 PTA of each bull in the latter group, which was based on information from its progeny, was compared with the traditional PA and the genomic PTA computed from 2004 data. The same process was repeated in the Jersey breed (1,361 older animals and 388 young bulls) and the Brown Swiss breed (512 older animals and 150 young bulls). Results in Table IV show the increase in reliability (**REL**) due to genomic information, as compared with the REL from parent average information only.

Trait	Increased reliability due to inclusion Genomic data		
	Holstein	Jersey	Brown Swiss
Net merit	+24	+8	+9
Milk yield	+26	+6	+17
Fat yield	+32	+11	+10
Protein yield	+24	+2	+14
Fat percentage	+50	+36	+8
Protein percentage	+38	+29	+10
Productive life	+32	+7	+12
Somatic cell score	+23	+3	+17
Daughter pregnancy rate	+28	+7	+18
Final classical score	+20	+2	+5
Udder depth	+37	+20	+8
Foot angel	+25	+11	-1

Table 4. Reliability changes due to the inclusion of genomic data in national genetic evaluations VanRaden *et al.*,(2009).

2.3.1.2 Beef animals

Recent advancements in sequencing and genotyping technologies have enabled a rapid evolution in methods for beef cattle selection. New tools will become available for beef producers to implement in the endeavor to efficiently produce high quality beef for today's consumer. Tools such as high-density genotyping assays and next generation sequencing instruments will help to shorten the generation interval, aid in the identification of causal mutations, increase the accuracy of EPDs on young sires and dams, provide information on gene expression and enhance our understanding of epigenetic and gut microbiome effects on cattle phenotypes. The ultimate objective of improving beef quality either by breeding or rearing factors. For genetic purposes, polymorphisms in some key genes have been reported for their association with beef quality traits. The sequencing of the bovine genome has dramatically increased the number of available gene polymorphisms. The association of these new polymorphisms with the variability in beef quality (e.g. tenderness, marbling) for different breeds in different rearing systems will be a very important issue. For rearing purposes, global gene expression profiling at the mRNA or protein level has already shown that previously unsuspected genes may be associated either with muscle development or growth, and may lead to the development of new molecular indicators of tenderness or marbling. Some of these genes are specifically regulated by genetic and nutritional factors or differ between different beef cuts (Hocquette *et al* ., 2007).

An Example for meat trait, a visibly distinct muscular hypertrophy (mh), commonly known as double muscling, occurs with high frequency in the Belgian Blue and Piedmontese cattle breeds. The autosomal recessive mh locus causing double-muscling condition in these cattle maps to bovine chromosome 2 within the same interval as myostatin, a member of the TGF-β super family of genes. Because targeted disruption of myostatin in mice results in a muscular phenotype very similar to that seen in double-muscled cattle, Kambadur *et al* ., (1997) have evaluated this gene as a candidate gene for double-muscling condition by cloning the bovine myostatin cDNA and examining the expression pattern and sequence of the gene in normal and double-muscled cattle. The analysis demonstrated that the levels and timing of expression do not appear to differ between Belgian Blue and normal animals, as both classes show expression initiating during fetal development and being maintained in adult muscle. Moreover, sequence analysis reveals mutations in heavy-muscled cattle of both breeds.

2.3.2 Goat

The International Goat Genome Consortium (IGGC) was created in March 2010 in Shenzhen and is coordinated by Professor Wen Wang, Zhang Wenguang (China) and Gwenola Tosser-Klopp (INRA). The initial activity of the consortium has focused on coordination in three areas, the results of which should benefit the planned activity in 3SR, namely: de novo goat genome assembly and resequencing; development of a Goat Radiation Hybrid Panel and Mapping; and production of a high-density SNP chip. Assembly of the goat genome is now underway in the Beijing Genome Institute (BGI). The average depth of sequencing is around 60X and optical mapping has been used to increase the scaffold size. Collection of samples for resequencing from around the world is now underway and once completed may provide information on domestication events (Zhang, 2011).

2.3.3 Sheep

The recent availability of the OvineSNP50 Beadchip (Illumina) opens promising perspectives on using molecular information in the management of breeding schemes of dairy and meat sheep populations. The large amount of information provided by this tool may have a strong impact on studies aimed at verifying the feasibility of Marker (or Gene) Assisted Selection or Genome-Wide Selection programs. The OvineSNP50 Beadchip has been recently used by the International Sheep Genomics Consortium (ISGC; *www.sheephapmap.org*) to genotype samples from 64 different sheep breeds. The sheep breed population may have great advantages by introducing molecular information in the selection scheme. Indeed the possibility to predict breeding values by genomic data might lead to reduce costs of phenotype recording and increase the number of selection objectives (Usai *et al.*, 2010).

2.3.4 Pig

The draft sequence, which is about 98% complete, allows researchers to pinpoint genes that are useful to pork production or are involved in immunity or other important physiological processes. It has the potential to increase genetic progress as well as offering an insight into pig disease and immunity traits and will assist in efforts to preserve the global heritage of rare, endangered and wild pigs. It also will be important for the study of human health because pigs are very similar to humans in their physiology, behavior and nutritional needs. To date, researchers have identified several genes or DNA regions that are associated with traits of economic importance including reproduction, growth, lean and fat quantity, meat quality traits and disease resistance. A number of gene and marker tests are now available commercially from genotyping service companies. Examples are CAST (meat quality), ESR and EPOR (litter size), FUT1 (E. coli disease resistance), HAL (halothane – meat quality, stress), IGF2 (carcase), MC4R (growth and fat), PRKAG3 (meat quality), RN (meat quality) (Walters, 2011).

2.3.5 Horse

The equine genome map was completed in the year 2007. Its applications are comparable to the development of antibiotics. Antibiotics benefited the health of everything; this is going to benefit the health of horses to the same extent in the next 50 years because of the things we can find out .More than 45 years ago, the introduction of antibiotics for use in animals greatly improved **animal health** and productivity. Today, **genetic testing**, gene therapy and the identification of **genetic markers** for certain diseases offer an even bigger opportunity for advancement in the equine industry (Steffanus, 2010).The Equine Genetic Diversity Consortium (EGDC) represents a collaborative, international community of equine researchers who are working to build a comprehensive understanding of genetic diversity among equine populations across the world and their work in future going to stimulate new studies into the origins of breeds and breed-defining traits and guide efforts to preserve genetic diversity (Petersen *et al* .,2011).

2.3.6 Poultry

Advances in DNA sequencing technology, the chicken genome project results, and a lot of hard work by researchers and geneticists are being combined by poultry primary breeders to change the way selection is done in pedigree flocks. Research projects involving pedigree lines from poultry primary breeders are demonstrating that genomic selection can increase the rate of progress in all traits, including economically important ones. Chickens have hundreds of thousands of SNPs in their genome. Phenotype information like growth rate, feed conversion in Broilers and egg production, egg quality and life span of the birds in layers are gathered from individuals within pedigree populations, and this is correlated with the individuals' SNPs. With the use of statistics, an individual's SNPs can be compared to information in the database and a genomic breeding value can be assigned to the animal. Analysis of an individual's SNPs can allow selection to take place earlier in the life cycle of the animal, thus genomic selection can cut the generation time down, which speeds the rate of progress (Keefe, 2011).

2.3.7 Buffalo

The water buffalo is vital to the lives of small farmers and to the economy of many countries worldwide. Not only are they draught animals, but water buffalo also produce meat, horns, skin and particularly the rich and precious milk that gives creams, butter, yogurt and many cheeses. Just recently, we sequenced a male water buffalo animal using the Illumina Genome

Analyzer II with a paired end of 101 bp. A full run generated more than 230 million reads, which resulted in approximately 46 Gb of high quality sequences. In this pilot study, we tested 210 pairs of water buffalo sequences for their bovine orthologs using a cross species mega BLAST approach developed at NCBI. Among these 210 pairs of sequences, only 7 pairs (3.3%) hit absolutely nothing in the bovine genome. One hundred twenty pairs of sequences (57%) had sole unique hits with one or both ends. For the remaining 83 pairs that had multiple hits in the bovine genome, 31 unique hits can be identified manually based on the aligned length and sequence similarity to the bovine orthologs. The alignment size between both species ranged from 42 bp to 122 bp, but with 93% alignments having more than 85 bp in length. This pilot study provides initial evidence that de novo water buffalo genome sequences can be comparatively assembled based on the cattle genome assembly (Jiang *et al* ., 2011).

3. Conclusions

The genetic improvement of animals is a fundamental, incessant, and complex process. In recent years many methods have been developed and tested. The genetic polymorphism at the DNA sequence level has provided a large number of markers and revealed potential utility of application in animal breeding. The invention of polymerase chain reaction (PCR) in accordance with the constantly increasing accuracy in DNA sequencing methods also represents a milestone in this endeavor. Selection of markers for different applications are influenced by certain factors - the degree of polymorphism, the automation of the analysis, radioisotopes used, reproducibility of the technique, and the cost involved. Presently, the huge development of molecular markers by DNA sequencing will continue in the near future. It is expected that molecular markers will serve as an underlying tool to geneticists and breeders to create animals as desired and needed by the society.

Farm animal genomic sequencing continues to attract audiences excited by the multitude of applications. The dairy and meat industry can now use cow and chicken genomic data to confirm the quality of dairy and meat products. Many scientists are using genomic information to determine disease-resistant genes in Cattle, Sheep, Goat and Horse and then are selectively mating the animals in order to create disease resistant animals. Many of the immediate practical applications of farm animal genome sequencing show potential for growth in this field. Recent technological advances in sequencing technologies and their corresponding bioinformatics tools will greatly aid a systematic analysis of molecular changes during physiological and pathological processes. This will provide insights into the biological mechanisms underlying animal health and allow better definition of traits with previously low heritability. Applying the 'genetic genomics' approach to farm animals is a major challenge, principally because maintaining sufficient numbers of large animals under standardized environmental conditions presents problems associated with management, housing and economics. Identifying QTL-affecting molecular profiles through these sequencing technologies may prove valuable when using marker-assisted selection to improve fertility, health and longevity in farm animals. In turn, molecular profiles of relevant tissues may also be useful to predict the consequences of selection before the animals reach an age at which classical phenotypic traits can be recorded. Similar approaches are currently being developed and tested in simple model organisms, but may also be applicable to laboratory and farm animals. Thus, Sequencing of farm animals is expected to have a major positive impact on sustainable livestock production.

4. References

Adam,D. (2002). Draft cow genome heads the field. *Nature*, Vol 417, pp 778.

Antin,P.B. and Konieczka,J.H., (2005).Genomic resources for chicken. *Developmental Dynamics*, Vol 232, pp 877–882.

Bentley, D.R., Balasubramanian, S., Swerdlow, H.P., Smith, G.P., Milton, J. and Brown, C.G., *et al.*, (2008). Accurate whole human genome sequencing using reversible terminator chemistry. *Nature*, Vol 456, pp 53–59.

Blakesley,R.W., Hansen,N.F., Mullikin,J.C., Thomas,P.J., McDowell,J.C., Maskeri,B., Young,A.C., Benjamin,B., Brooks,S.Y., Coleman,B.I. et al. (2004) An intermediate grade of finished genomic sequence suitable for comparative analyses. *Genome Research*, Vol 14, pp 2235–2244.

Bleck, G.T., Conroy, J.C. and Wheeler, M.B., (1996). Polymorphisms in the bovine ß-casein 5' flanking region, *Journal of Dairy Science*, Vol 79, pp 347–349.

Cockett,N.E., Shay,T.L. and Smit,M. (2001). Analysis of the sheep genome. *Physiological. Genomics*, Vol 7, pp 69–78.

Cole, J. B., VanRaden, P.M., O'Connell, J.R., Van Tassell, C.P., Sonstegard,T.S., Schnabel,R.D., Taylor,J.F. and Wiggans, G.R., (2009). Distribution and location of genetic effects for dairy traits. *Journal of Dairy Science*, Vol 92, pp 2931-2946.

Da, Y., (2003). Statistical analysis and experimental design for mapping genes of complex traits in domestic animals. *Bioinformatics*, Vol 30, No12, pp 1183–1192.

Dequeant,M.L. and Pourquie,O., (2005). Chicken genome: new tools and concepts. *Developmental Dynamics*, Vol 232, pp 883–886.

Eggen, A. (2008). Whole genome sequencing in livestock species: The end of the beginning? European Animal Disease Genomics Network of Excellence for Animal Health and Food *Safety Animal Disease Genomics: Opportunities and Applications* 10th - 11th June 2008 , Edinburgh, UK.

Fadiel, A., Anidi, I. and Eichenbaum, K. D.,(2005). Farm animal genomics and informatics: an update. *Nucleic Acids Research,* Vol 33, No 19, pp 6308–6318.

Fields, C., Adams, M. D., White, O., Venter, J. C. (1994). How many genes in the human genome. *Nature Genetics,*Vol 7, pp 345–346.

Harismendy, O., Ng, P.C., Strausberg, R.L., Wang, X., Stockwell, T.B., Beeson, K.Y., Schork N.J., Murray, S.S., Topol, E.J., Levy, S. and Frazer, K.A.,(2009). Evaluation of next generation sequencing platforms for population targeted sequencing studies. *Genome Biology,*Vol 10, pp 32.

Hocquette, J. F., Lehnert, S., Barendse, W., Cassar-Malek, I. and Picard, B.,(2007). Recent advances in cattle functional genomics and their application to beef quality. *Animal,* Vol 1, pp 159–173.

Jeffreys, A.J., Wilson V. and Thein S.L., (1985). Individual-specific fingerprint of human DNA. *Nature,*Vol 316, pp 76-79.

Jiang, Z. and Ott, T.L., (2010).Quantitative Genomics of Reproduction. *Reproductive Genomics in Domestic Animals*. Cockett, N.E., pp 1-7. Wiley and Blackwell publishers, July 2010. ISBN: 978-0-470-96182-7, Iowa USA.

Jiang, Z., Michal, J. J., Zhang, M., Zambrano-Varon, J. and Ealy, A. D., (2011). Whole Genome Sequencing Of Water Buffalo – Initial Characterization For Developing Novel Assembly Strategy. Goat Genome Sequencing and Its Annotation. *Plant & Animal Genomes XIX Conference* , San Diego, CA January 15-19. W533 Cattle/sheep.

Johnston, B. T. D. J. and Graser, H.U.,(2009). Integration of Dna markers into breedplan EBVs Proceedings of the Eighteenth Conference *Matching Genetics and the Environment a New Look at and Old Topic. Proceedings of Association Advancement of Animal Breeding and. Genetics,* 2009. Vol 18,pp 30-33.

Kambadur,R., Sharma, M., Smith, T.P.L. and Bass,j. J.(1997). Mutations in myostatin (GDF8) in Double-Muscled Belgian Blue and Piedmontese Cattle. *Genome research,* Vol 7, No1,pp 910-915.

Keefe, T. O., (2011). Poultry breeders see paradigm shift with genomic selection. More rapid improvement of economically important traits may be just around the corner for the egg layer and broiler industries. *Egg industry,* Vol 116, No 8, pp 8-11.

Kerstens, H.H.D., (2010). Bioinformatics approaches to detect genetic variation in whole genome sequencing data. PhD thesis submitted to *Wageningen University.*

Klymiuk,N. and Aigner,B., (2005). Reliable classification and recombination analysis of porcine endogenous retroviruses. *Virus Genes,* Vol 3, pp 357–362.

Koczan, D., Hobom, G.and Seyfert, H.M., (1991). Genomic organization of the bovine alpha-s1 casein gene. *Nucleic Acid Research,*Vol 19, pp 5591-5596.

Koopaei, H. K. and Koshkoiyeh, A. E. (2011).Application of genomic technologies to the improvement of meat quality in farm animals *Biotechnology and Molecular Biology Review.* Vol 6, No6, pp, 126-132.

Kumar,S. and Hedges,S.B., (1998).A molecular timescale for vertebrate evolution. *Nature,* Vol 392, pp 917–920.

Larkin,D.M., Everts-van der Wind,A., Rebeiz,M., Schweitzer,P.A., Bachman,S., Green,C., Wright,C.L., Campos,E.J., Benson,L.D., Edwards,J. *et al.* (2003). Acattle–human comparative map built with cattle BAC-ends and human genome sequence. *Genome Research,* Vol 13, pp 1966-1972.

Leif Andersson,(2001). Genetic dissection of phenotypic diversity in farm animals, *Nature review,* Vol 2, pp 130-138.

Li, H., Ruan, J., and Durbin, R., (2008). Mapping short DNA sequencing reads and calling variants using mapping quality scores. *Genome Research,* Vol 18, pp 1851-1858.

Li, R., Fan, W., Tian, G., Zhu, H., He, L., Cai, J., Huang, Q. and Cai, Q., *et al* .,(2009).The sequence and de novo assembly of the giant panda genome. *Nature,* Vol 463, 311-317.

Liu, x., Zhang, H., Li, H., Li, N., Zhang, Y., Zhang, Q., Wang, S., Wang, Q. and Wang, H., (2008). Fine mapping quantitative trait loci for body weight and abdominal fat traits: Effects of marker density and sample size. *Poultry Science,* Vol 87, pp 1314-1319.

Loftus, R.T., MacHugh, D. E., Bradley, D. G., Sharp, P. M. & Cunningham,P.,(1994).Evidence for two independent domestications of cattle. *Proceedings of National Academy of Sciences. USA,Vol* 91, pp 2757–2761.

Manfredi, E., Ricordeau, G., Barbieri, M.E., Amigues, Y. and Bibe, B., (1995). Genotype caseines αs1 et selection des boucs sur descendance dans les races Alpine et Saanen. *Genetics Selection and Evolution,* Vol 27, pp 451-458.

Martin, P., Szymanowska, M., Zwierzchowski, L. and Leroux, C.,(2002). The impact of genetic polymorphisms on the protein composition of ruminant milks. *Reproduction. Nutrition and Development* , Vol 42, pp 433-459.

Montaldo, H.H. and Meza-Herrera, C.A., (1998). Use of molecular markers and major genes in the genetic improvement of livestock. *Journal of Biotechnology. Vol* 1, No 2, pp 1-7.

Mouchel,N., Tebbutt,S.J., Broackes-Carter,F.C., Sahota,V., Summerfield,T., Gregory,D.J. and Harris,A. (2001), The sheep genome contributes to localization of control elements in a human gene with complex regulatory mechanisms. *Genomics*, Vol 76, pp 9–13.

Novaes, E., Drost, D.R, Farmerie, W.G., Pappas, G.J., Grattapaglia, D., Sederoff, R.R., and Kirst, M.(2008). High-throughput gene and SNP discovery in Eucalyptus grandis, an uncharacterized genome. *BMC Genomics,*Vol 9, pp 312.

Petersen, J. L., Mickelson, J. R., Andersson, L. S., Bailey, E., Bannasch, D. L. and and Binns ,M. M. *et al.*, (2011). The Equine Genetic Diversity Consortium: An International Collaboration To Describe Genetic Variation In Modern Horse Breeds. *Plant & Animal Genomes XIX Conference*, San Diego, CA January 15-19. P 617: Equine

Perez-Enciso ,M. and Ferretti, L., (2010). Massive parallel sequencing in animal genetics: wherefroms and Wheretos. *Animal Genetics,*Vol 41, pp 561–569.

Pushkarev, D., Neff, N.F. and Quake, S.R., (2009). Single-molecule sequencing of an individual human genome. *Nature Biotechnology,*Vol 27, 847-852.

Rempel, W.E., Lu, M., El-Kandelgy, S., Kennedy, C.F., Irvin, L.R., Mickelson, J.R. and Louis, C.F., (1993). Relative accuracy of the halothane challenge test and a molecular genetic test in detecting the gene for porcine stress syndrome. *Journal of Animal Science,* Vol 71, pp 1395-1399.

Rothschild, M. F. and Soller, M. (1997). Candidate gene analysis to detect genes controlling traits of economic importance in domestic livestock. *Probe* , Vol 8,pp 13–22.

Rothschild,M.F. (2003). From a sow's ear to a silk purse: real progress in porcine genomics. Cytogenet. Genome Res., 102, 95–99.

Rothschild,M.F., (2004). Porcine genomics delivers new tools and results: this little piggy did more than just go to market. *Genetic Research*, Vol 83,pp 1–6.

Schild, T.A., Wagner, V. and Geldermann, H., (1994). Variants within the 5' - flanking regions of bovine milk protein genes: kappa -casein gene. *Theoretical and Applied Genetics*, Vol 89,pp 116-120.

Shuster, D.E., Kehrli Jr, M.E., Ackermann, M.R. and Gilbert, R.O., (1992). Identification and prevalence of a genetic defect that causes leukocyte adhesion deficiency in Holstein cattle. *Proceedings of National Academy of Sciences. USA*, Vol 89, pp 9225-9229.

Steffanus, D. (2010). Genetics: The new frontier .The equine genome map is as important to the future of horses as the development of antibiotics was 45 years ago. America's Horse Daily. March 11.

Tabor, H. K., Risch, N.J. and Myers, R.M., (2002). Candidate-gene approaches for studying complex genetic traits: Practical considerations. *Nature Reviews Genetics*, Vol 3: 391-397.

Walsh, B. (2000). Minireview: Quantitative genetics in the age of genomics. *Theoretical. Population Biology*, Vol 59, pp 175-184.

Walters, R., (2011). More commercial benefits on horizon as pig genome project nears completion. Information from pig genome already being used in the industry. *Pig International* ,Vol 41, No2: pp 16.

Wang, J, Wang, W., Li, R., Li, Y., Tian, G., Goodman, L., et al., (2008). The diploid genome sequence of an Asian individual. *Nature*, Vol 456, pp 60-65.

Wernersson,R., Schierup,M.H., Jorgensen,F.G., Gorodkin,J., Panitz,F., Staerfeldt,H.H., Christensen,O.F., Mailund,T., Hornshoj,H., Klein,A. *et al.*, (2005). Pigs in sequence space: a 0.66X coverage pig genome survey based on shotgun sequencing. *BMC Genomics*,Vol 6,pp 70.

Wheeler, D.A., Srinivasan, M., Egholm, M., Shen. Y., Chen, L., McGuire, A., *et al* .,(2008). The complete genome of an individual by massively parallel DNA sequencing. *Nature*, Vol 452, pp 872-876.

Wiedmann, R.T., Smith, T.P.L., Nonneman, D.J., (2008). SNP discovery in swine by reduced representation and high throughput pyrosequencing. *BMC Genomics* ,Vol 9,pp 81.

Usai, M. G., Sechi, T., Salaris, S., Cubeddu, T., Roggio, T., Casu, S. and Carta, A.,(2010). Analysis of a representative sample of Sarda breed artificial insemination rams with the OvineSNP50 BeadChip. *Proceedings ICAR 37th Annual Meeting* – Riga, Latvia (31 May - 4 June, 2010), pp 7-10.

VanRaden, P. M., Van Tassell, C.P., Wiggans,G.R., Sonstegard,T.S., Schnabel,R.D., Taylor, J.F. and Schenkel, F.,(2009). Reliability of genomic predictions for North American Holstein bulls. *Journal of Dairy Science*, Vol 92, pp 16-24.

VanRaden, P., Connell, J. R. O., Wiggans, G. R. and Weigel, K.A.,(2011). Genomic evaluations with many more genotypes. *Genetics Selection Evolution*,Vol 43, pp1-11.

Van Tassell, C. P., Smith, T. P. L., Matukumalli, L. K., Taylor, J. F., Schnabel, R. D., C Lawley, C.T., Haudenschild,C.D., Moore,S.S., Warren,W.C., and Sonstegard,T.S.,(2008). SNP discovery and allele frequency estimation by deep sequencing of reduced representation libraries. *Nature Methods*, Vol 5, pp247–252.

Yahyaoui, M. H., (2003). Genetic polymorphism in goat, Study of the kappa casein,beta lactoglobulin, and stearoyl coenzyme A desaturase genes. Ph.D thesis submitted to *Universidad Autónoma de Barcelona.*, Bellaterra, 2003.

Yan, H., Kinzler, K.W. and Volgelstein, B., (2000). Genetic testing, present and future. *Science*, Vol 289, 1890-1892.

Zhang, W. , Sayre, B. L., Xu, X., Tosser, G. , Li, J. and Wan, W., (2011). Goat genome sequencing and its annotation. *Plant & Animal Genomes XIX Conference* San Diego, CA January 15-19. W135 Cattle/sheep.

Zhu, M. and Zhao, S., (2007). Candidate gene identification approach: Progress and challenges. *International Journal of Biological Sciences*, Vol 3, No 7, pp 420-427.

7

Nucleic Acid Aptamers as Molecular Tags for Omics Analyses Involving Sequencing

Masayasu Kuwahara[1] and Naoki Sugimoto[2]
[1]Chemistry Laboratory of Artificial Biomolecules (CLAB),
Graduate School of Engineering, Gunma University
[2]Frontier Institute for Biomolecular Engineering Research (FIBER) and
Faculty of Frontiers of Innovative Research in Science and Technology (FIRST),
Konan University
Japan

1. Introduction

Since the beginning of the $1,000 genome project, sequencing technologies have progressed rapidly and equipments capable of massively parallel sequencing have been developed. These machines can perform 3G DNA sequencing corresponding to the entire human genome within only a few hours. Indeed, using such machines, quantitative analysis methods for multiple types of RNA, even in a single cell, are very close to being developed. This would be useful for transcriptome studies including mRNA, rRNA, tRNA, and other non-coding RNA. However, these methods would not be applicable to proteome and other omics analyses because non-nucleic acid molecules, such as proteins and peptides, do not have nucleotide sequences. If non-nucleic acid molecules could be converted to a nucleotide sequence, analytical methods that use massively parallel sequencing could also be applied to omics analyses for non-nucleic acid molecules. Nucleic acid aptamers (specific binders) can be readily amplified by polymerase reaction and decoded by sequencing. Therefore, it is possible to apply them as molecular tags to quantitative biomolecular analyses and single cell analyses, among others. This chapter focuses on the development of nucleic acid aptamers and the outlook for related technologies.

2. Construction of random screening systems

Random screening refers to a methodology for obtaining molecules that perform a desired activity, such as catalysis and specific binding, by displaying masses of diverse molecules called a library and selecting from them. This method involves phage display [Smith & Petrenko, 1997] and split–mix combichem [Geysen *et al.*, 2003] as well as *in vitro* selection [Robertson & Joyce, 1990; Ellington & Szostak, 1990; Tuerk & Gold, 1990], *i.e.*, it is a random screening method for functional nucleic acids.

2.1 Beginning of screening functional nucleic acids

Functions of nucleic acids, such as DNA and RNA, have been considered to be only to preserve, transfer, and express genetic information. The prevailing notion is that proteins are functional molecules that play an important role in the body and that DNA is the blueprint of proteins and RNA is the mediator. Therefore, the potential of nucleic acids to behave as functional molecules has not attracted much interest. However, since the 1980s, with the discovery by Cech *et al.* and Altman et al. of RNA enzymes (ribozymes) that catalyze RNA self-cleavage or RNA transesterification in splicing [Zaug & Cech, 1986; Guerrier-Takada *et al.*, 1983], much interest has been focused on the creation and application of functional nucleic acids with new activities that are different from the typical activities of nucleic acids.

A screening methodology to select functional RNA molecules that can catalyze a specific reaction (ribozyme) or bind to a specific molecule (aptamer) from the RNA library (RNA pool of miscellaneous random sequences) was independently reported by Szostak *et al.*, Joyce *et al.*, and Gold et al. around 1990. This random screening method for functional nucleic acids is called *in vitro* selection. This is often referred to as SELEX (systematic evolution of ligands by exponential enrichment), particularly when used for screening aptamers. Using the *in vitro* selection method, successful screenings of functional RNA were demonstrated, followed by the creation of the DNA enzyme as a catalyst and the DNA aptamer as a specific binder [Breaker & Joyce, 1994].

2.2 Random screening of nucleic acid aptamers

Single-stranded RNA and DNA (ssRNA and ssDNA, respectively) with a particular sequence, which can be selected by the SELEX method, exert the aptamer activities by forming a specific steric structure with intramolecular hydrogen bonding, stacking interactions, electrostatic interactions, and metal coordination. Using the SELEX method, various RNA/DNA aptamers, which specifically bind to a broad spectrum of targets involving small molecules such as nucleotides or amino acids, vitamins, and antibiotics; macromolecules such as thrombin; and particles such as Rous sarcoma virus (RSV), have been created artificially [Uphoff *et al.*, 1996]. Since the discovery of the riboswitch early in the 21st century, it has been known that nucleic acid aptamers also exist in nature [Winkler *et al.*, 2002].

In the SELEX method, the screening of nucleic acid aptamer is performed using an ssDNA pool of miscellaneous random sequences as the initial library. Oligonucleotide containing approximately 10^{13-15} diversities of sequences is chemically synthesized. Typically, its length is approximately forty mer to one hundred and several tens mer, though the length can vary depending on the size of the target and the screening system that the researchers want to construct. In the case of DNA aptamer screening (Fig. 1), the initial library is applied to solid support, such as sepharose gel and nitrocellulose membrane, such that the target is immobilized and active species are separated from nonactive ones. The DNA with target activity is amplified by polymerase chain reaction (PCR) to give the corresponding dsDNA, and then ssDNA is prepared by an appropriate treatment of the complementary strand, such as degradation using λ exonuclease and biotin trapping using avidin gels. The resulting ssDNA library is used for the next round, and repeated cycles of separation and amplification lead to enriched sequences that show affinity to the target. The DNA aptamers are individually isolated from the enriched pool using a cloning method, and their sequences and binding properties to the target are analyzed. In the case of RNA aptamer screening, two processes are involved. The initial synthetic DNA library is converted to a RNA library using RNA polymerase and after affinity separation, the active RNA is converted and amplified to DNA using a reverse transcription PCR (RT-PCR); this library is subjected to the previously described DNA aptamer screening scheme.

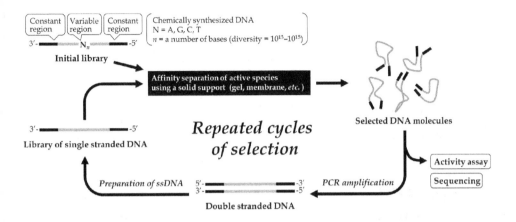

Fig. 1. General scheme of the SELEX method using the DNA library.

3. Chemical modification

Many examples of nucleic acid aptamers have been reported. However, some defects have been revealed. A major defect is physical instability due to degradation by endogenous nucleases. Unmodified nucleic acid aptamers are immediately degraded *in vivo*. Furthermore, a limitation on the binding affinity and specificity are anticipated because nucleic acids consist of a combination of only four nucleotides whereas proteins consist of a combination of twenty amino acids with diverse functional groups in their side chains. Therefore, the development of an advanced screening method to obtain modified nucleic acid aptamers from chemically modified libraries, in which different functionalities that cannot be found in a nucleotide are incorporated, has been performed.

3.1 Polymerase reaction involving modified nucleotides

A key step in the SELEX method is the amplification of selected nucleic acid molecules by affinity separation. This method would make good use of a unique feature of the DNA molecule: it can be amplified and replicated through PCR. Other organic molecules do not possess such a feature. In order to adapt the method to modified nucleic acids as well as to natural RNA and DNA, modified nucleic acids should be enzymatically synthesized from a natural DNA template by polymerase reactions with high efficiency. Conversely, DNA should also be efficiently produced from a modified nucleic acid template. Of course, those conversions are not necessary if the modified nucleic acid used could be directly amplified by PCR in the same manner as DNA.

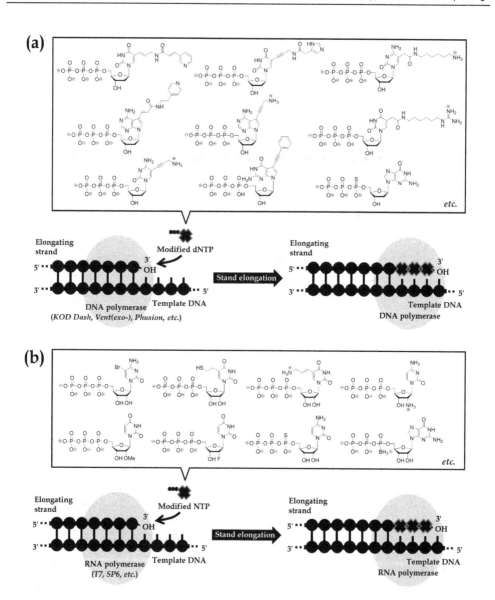

Fig. 2. Enzymatic production of modified DNA (a) and modified RNA (b).

Various types of modified nucleic acids have been designed and polymerase reactions using them have been reported. A monomer unit of nucleic acid consists of a base, sugar, and phosphate moieties, and each component could be an object of chemical modification. When modified nucleic acids are produced from a DNA template, modified nucleoside triphosphate can be used instead of the corresponding natural nucleoside triphosphate as a substrate for the polymerase reaction (Fig. 2a). If the modified nucleoside triphosphate acts as a good substrate for the polymerase reaction, the corresponding modified nucleic acids could be efficiently produced. The modification of the sugar and phosphate moieties would tend to decrease the efficiency of the polymerase reaction to a much greater extent than the base moiety does, although this depends on the type of chemical modification [Kuwahara *et al.*, 2006; 2008; 2009]. In particular, 2'-deoxynucleoside-5'-triphosphate analogs with pyrimidine substituted at the 5th position and purine substituted at the 7th or 8th positions of the base moiety tend to be acceptable for DNA polymerase and act as good substrates [Sakthivel & Barbas, 1998; Lee *et al.*, 2001; Tasara *et al.*, 2003]. Furthermore, α-phosphate analogs, where oxygen is replaced with other chemical species, can also be acceptable for DNA polymerase [Andreola *et al.*, 2000]. Analogs with amino (-NH$_2$), fluoro (-F), and methoxy (-OMe) groups substituted at the 2' position of the sugar moiety are known to work in the RNA polymerase reaction [Aurup *et al.*, 1992; Kujau & Wölfl, 1998].

The chemical structure and replaced position of the modified group as well as the type of polymerase used would affect the production of modified nucleic acids. Among the commercially available thermostable DNA polymerases used for PCR, those belonging to the evolutional family B, such as *KOD Dash*, *Vent(exo-)*, and *Phusion*, are thought to be suitable for modified nucleic acid production. In particular, *KOD Dash* and mutants of *KOD* DNA polymerases derived from *Thermococcus kodakaraensis* would be the most suitable for modified DNA synthesis as long as misincorporation rates and successive incorporation of modified substrates are investigated. In contrast, *Taq* or *Tth* DNA polymerases, which belong to the evolutional family A, have been found to be sensitive to chemical modification and not suitable for modified DNA synthesis. In PCR, efficient incorporation of modified substrates into the extending strand is required and the modified nucleotide strands produced should act as templates for the next thermal cycle. Though this double barrier make modified DNA synthesis difficult, some 2'-deoxynucleoside-5'-triphosphate analogs of C5-substituted pyrimidine, C7-substituted 7-deaza-purine, and α-phosphoro-thioate have been found to be good substrates for PCR when combined with an appropriate DNA polymerase.

Modification of nucleotides often decreases the reaction efficiency and results in small or reduced amounts of the modified DNA with PCR; the modification would cause sequence dependency on the reaction rate, resulting in a bias in the sequences that emerged from the screening. Indeed, kinetic studies on a polymerase reaction using modified nucleotides showed that the reaction with successive incorporations of modified nucleotides was found to be far more inefficient than the reaction with natural nucleotides. In the SELEX using modified DNA, PCR amplification is performed using natural triphosphates in many cases to circumvent this problem. First, the modified DNA is prepared by a primer extension reaction (PEX) using the modified substrate and natural DNA template. After affinity separation, the natural DNA is synthesized and amplified from the selected modified DNA as a template during PCR using natural triphosphates. Then, the modified DNA for the next selection round can be prepared by transcribing the amplified DNA using PEX. This scheme does not contain the simultaneous use of modified triphosphate substrate and modified DNA template. Therefore, the modified DNA, in which many kinds of functionalities are incorporated, can be synthesized more efficiently compared with the previously described method of direct PCR amplification of modified DNA. Indeed, this has enabled SELEX to be used with libraries of doubly or triply modified DNA prepared using PEX [Perrin *et al.*, 2001; Sidorov *et al.*, 2004; Hollenstein *et al.*, 2009].

For enzymatic synthesis of modified RNA, the template DNA is transcribed using modified ribonucleoside triphosphates instead of natural triphosphates. Fig. 2b shows modified analogs accepted as substrates by RNA polymerase, e.g., T7 RNA polymerase and SP6 RNA polymerase. Modified RNA, particularly that prepared using a uridine analog with various functionalities at the 5th position or uridine/cytidine analogs with a fluorine group or an amino group at the 2' position, is often applied to the SELEX method. Although the incorporation efficiency of modified analogs is inferior to that of natural substrates, they can provide a relatively long strand of modified RNA as a full-length product corresponding to template DNA. In the SELEX method using modified RNA, after affinity separation, the selected modified RNA is reverse transcribed to DNA and then amplified by PCR; natural nucleoside triphosphates are used as substrates in these processes. The modified RNA for the next selection round is prepared by transcribing the amplified DNA. In reverse transcription, AMV (avian myeloblastosis virus) reverse transcriptase, Tth DNA polymerase with reverse transcription activity, or other types of modified RNA polymerases are often used. The aforementioned modified RNAs were found to act as templates for the reverse transcription catalyzed by these polymerases to yield the corresponding DNA strands.

3.2 Modified nucleic acid aptamers

To use nucleic acid aptamers in vivo, chemical modification is indispensable. The first aptamer drug, "Macugen," which is used to treat age-related macular degeneration (AMD), comprises modified RNA including 2'-fluoropyrimidine nucleotides (U, C) and 2'-methoxy purine nucleotides (A, G). This aptamer was created as follows: 1) Sequences that bound to vascular endothelial growth factor (VEGF) using a modified RNA library including 2'-fluoropyrimidine nucleotides were selected by SELEX; 2) Natural purine nucleotides were replaced with 2'-methoxy purine nucleotides, in part, only at sites where its binding affinity was retained after the replacement; 3) Chemical modifications of the 5' and 3' ends with branched polyethylene glycol strands and 3'-thymidylate were introduced [Ruckman et al., 1998].

In step 1 of the aforementioned process, T7 RNA polymerase was used as an enzyme, and 2'-fluoropyrimidine nucleoside triphosphates and natural purine nucleoside triphosphates were used as substrates to prepare the modified RNA library. Here, modified purine nucleoside triphosphates could not be used due to the limited capability of T7 RNA polymerase in producing modified RNA. Steps 2 and 3 are called post-SELEX chemical modifications, which can increase nuclease resistance and extend the periods of hemodynamic stasis. In general, chemical modifications have been shown to remarkably enhance biostability and occasionally increase the binding capability of nucleic acid aptamers, as in the following examples.

A sugar-modified RNA aptamer specific to human neutrophil elastase (HNE) is highly stable in serum with a half-life of approximately 20 hours, while natural RNA is degraded within approximately 5 minutes [Lin et al., 1994]. This aptamer includes 2'-aminopyrimidine nucleotides (U, C) instead of the corresponding natural nucleotides. In addition, the improved nuclease resistance of modified RNA aptamers with other types of modifications, i.e., 2'-fluorine and 2'-methoxy groups, were experimentally confirmed. Furthermore, inserting only a single bridged nucleotide analog having 2'-CH(Ph)OCH$_2$-4' at the 3' end of a thrombin binding DNA aptamer (TBA) could increase the resistance to venom exonuclease up to approximately 30 times (Fig. 3) [Kasahara et al., 2010].

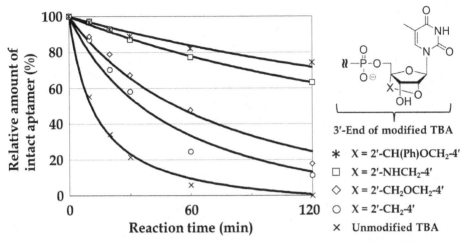

Fig. 3. The time course of the degradation of thrombin binding DNA aptamers (TBAs) by phosphodiesterase I; Total quantities of the products were set at 100% in each reaction mixture.

A base-modified RNA aptamer specific to the human immunodeficiency virus (HIV) Rev protein obtained from a library of modified RNA containing 5-iodouridine instead of uridine could be bound to the target with a somewhat higher binding affinity (K_d = 0.8 nM) compared with the corresponding natural RNA aptamer [Jensen et al., 1995]. The unique feature of this base-modified aptamer is to form a cross-link with the target protein by UV irradiation; halogenated uracil is known to form a covalent bond with a nearby electron-rich amino acid residue by photoirradiation.

Such photo-cross-linkable aptamers are called photoaptamers. They are generally created by photoSELEX, which involves a photo-cross-linking process in the conventional selection cycle. Photoaptamers, in particular, would have an advantage for use in protein assays because much more stringent washing of excess or non-specific proteins can be done compared with antibody-based sandwich assays, because photoaptamers can covalently and irreversibly bind to a target.

A few phosphate-modified RNA aptamers have been reported because the commonly used T7 RNA polymerase could accept limited triphosphate analogs modified at the phosphate moiety as substrates. Modified nucleoside triphosphates involving phosphorothioate and boranophosphate [Lato et al., 2002] are normally used as substrates. A phosphate-modified RNA aptamer specific for bFGF was obtained by screening the library of modified RNA exclusively involving phosphorothioate inter-nucleoside linkages, which could greatly enhance nuclease resistance [Jhaveri et al., 1998]. The aptamer bound to bFGF with a binding affinity of K_d=1.8nM which was approximately five times lower than that observed for a 2'-amino RNA aptamer for the same target (K_d = 0.35 nM). However, its strong nuclease resistance was confirmed.

Similarly, various modified DNA aptamers involving 2'-deoxynucleoside analogs with base or phosphate modifications have been selected by SELEX using modified DNA libraries. In both cases with modified RNA/DNA, the inefficiency of enzymatically modified nucleic acid production would occasionally become a bottleneck. Another methodological approach to create modified nucleic acid aptamers is a mirror-image aptamer coined Spiegelmer. To obtain Spiegelmers, there is no need to enzymatically prepare modified nucleic acid libraries during the selection cycles. Instead, screening is performed using natural RNA/DNA libraries by means of a mirror image of the target molecule. After screening, mirror-image aptamers with the same sequences as the selected aptamers were chemically synthesized using L-ribonucleotides or L-deoxyribonucleotides. Until date, Spiegelmers that bind specifically to L-arginine, D-adenosine, L-vasopressin, and others have been created; they were found to have improved nuclease resistance [Nolte et al., 1996].

4. Affinity and specificity

Affinities of well-known representative nucleic acid aptamers are shown in Fig. 4, along with those of specific protein binders, such as antibody, streptavidin, and rectin. In general, the binding affinity between molecules can be numerically expressed by the dissociation constant (K_d). At equilibrium for the association/dissociation reaction R + L $\rightleftarrows$ RL, K_d is defined by: K_d = [R][L]/[RL], where [R], [L], and [RL] are concentrations of free receptor, free ligand, and their interaction complex, respectively; smaller K_d values indicate higher binding affinities. For example, K_d values of biotin/streptavidin and digoxigenin/anti-digoxigenin antibody, which are often used as research reagents, are 4 × 10^{-14} (40 fM) and 1.1 × 10^{-8} (11 nM), respectively [Holmberg et al., 2005; Tetin et al., 2002]. Those for galactose/jacalin and mannose/concanavalin A are 1.6 × 10^{-5} (16 μM) and 2.0 × 10^{-4} (200 μM), respectively [Smith et al., 2003]. Immunoglobulin G (IgG) has K_d values of approximately 10^{-7} (100 nM) or less, which is classified as a high affinity antibody.

K_d values of protein binding aptamers specific for keratinocyte growth factor (KGF), vascular endothelial growth factor (VEGF), and thrombin are 3 × 10^{-13} (0.3 pM), 5 × 10^{-11} (50 pM), and 5 × 10^{-10} (500 pM), respectively; these have excellent binding affinities [Pagratis et al., 1997; Kubik et al., 1994; Ruckman et al., 1998]. The VEGF binding aptamer is Macugen. The K_d value of an anti-VEGF antibody, bevacizumab, which is used for cancer therapy, is 1.1 × 10^{-9} (1.1 nM). In general, the K_d value required for antibody drugs for molecular targeted therapies should be approximately 10^{-9} (1 nM) or less. Hence, nucleic acid aptamers comparable to antibodies in terms of binding affinity can be created by SELEX with sophisticated techniques as long as the target is a high molecular weight molecule like a protein.

Regarding nucleic acid aptamers specific for small molecules, the K_d value of a biotin binding aptamer, for example, is about 6 × 10^{-6} (6 μM), which is 10^8 or more times greater than that of streptavidin [Wilson et al., 1998]. Except for a few examples, the reported small molecule binding aptamers that have been artificially created have K_d values ranging from approximately 10^{-7} to 10^{-4} (several hundred nM to several hundred μM). To use these as research reagents and diagnostic agents, their binding affinities need to be increased up to at least 10^{-7} or less, hopefully 10^{-8} or less.

Ligand binding sites of riboswitches found in the mRNA of bacteria and the like are considered to be naturally occurring aptamers. Some of these aptamers exhibit very high binding affinities for their small molecule targets, and their K_d values are several nM. For example, K_d values of aptamers specific for S-adenosyl-methionine, flavin mono nucleoside, and guanine are 4 × 10^{-9} (4 nM), 5 × 10^{-9} (5 nM), and 5 × 10^{-9} (5 nM), respectively [Winkler et al., 2003; Winkler et al., 2002; Mandal et al., 2003]. Thus, it is possible that aptamers that bind strongly to small molecules could be created by further improvements in screening methods.

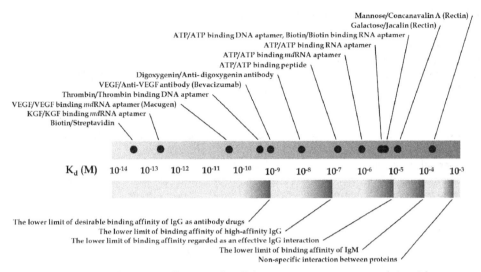

Fig. 4. Comparisons of binding affinities of well-known representative nucleic acid aptamers with those of proteins having specific binding abilities.

Effects of chemical modifications on binding affinity and specificity remain unclear till date. However, it has been suggested that chemical modifications that could constrain structural fluctuations would favorably influence binding affinity. This does not contradict that, in many cases, a target/nucleic acid aptamer complex can be stabilized with divalent metal cations like Mg^{2+} and Ca^{2+}. To characterize effects of chemical species in a library in terms of affinity and specificity, binding properties of aptamers specific for a target, which are respectively selected from libraries different in chemical species using a standardized common screening protocol, should be systematically analyzed.

In previous studies, ATP binding aptamers were obtained respectively from RNA, DNA, and modified RNA and modified DNA, although there were some differences in the screening protocol [Sazani et al., 2004; Huizenga & Szostak, 1995; Vaish et al., 2003; Battersby et al., 1999]. Modified RNA and modified DNA contain 5-(3-amino)propyl-uridine and 5-(3-amino)propyl-2'-deoxyuridine, respectively. K_d values of the obtained aptamers ranged from approximately10^{-7} to 10^{-6} (several hundred nM to several μM), and remarkable differences depending on chemical species in libraries were not observed. Incidentally, the K_d value of an ATP binding peptide comprised 62 amino acids obtained using mRNA display selection was 1.9×10^{-7} (190 nM); the binding affinity of this peptide is likely to be higher to some degree than those of the aforementioned nucleic acid aptamers [Chaput & Szostak, 2004].

Fig. 5 shows a comparison of the binding specificities of an RNA aptamer, a modified RNA aptamer, and a peptide that bind to ATP. The affinity of the RNA aptamer is 64-fold lower because of a lack of γ–phosphate in ATP, and is 1100-fold lower because of lack of both β– and γ–phosphates. Although its affinity is 600-fold lower because of a replacement of an adenine base with a pyrimidine base, its specificities for the base and sugar moieties in ATP tend to be less sensitive than that for the phosphate moiety. In contrast, the modified RNA aptamer accurately recognizes the base moiety, but its specificities for phosphate and sugar moieties are likely to be low. Base discernment of the ATP binding peptide is very high; it can accurately distinguish even hypoxanthine in ITP from adenine in ATP. It can recognize the sugar moiety much more clearly than the other two binding molecules, but it is inferior to the RNA aptamer with regard to recognizing the phosphate moiety. While the ATP binding peptide would recognize the target over its entire structure, the ATP binding nucleic acid aptamers also show high recognition capabilities, such as discriminating the detailed partial structures of the target molecule. Unfortunately, comparisons of RNA and modified RNA aptamers do not indicate any superiority owing to the introduction of chemically modified groups. However, (3-amino) propyl groups introduced were found to be essential in order that the modified RNA aptamer could bind to ATP, which indicated that the modified RNA aptamer could recognize the target molecule in a manner that was different from that of a natural RNA aptamer.

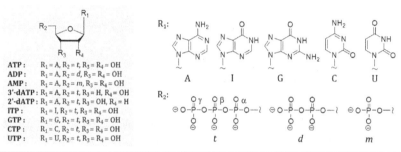

Recognition site	Phosphate moiety			Sugar moiety		Base moiety			
Target molecule	ATP	ADP	AMP	3'-dATP	2'-dATP	ITP	GTP	CTP	UTP
ATP binding RNA aptamer	1	64	1100	8.5	10	66	19	680	610
ATP binding *md*RNA aptamer	1	1	2.1	2.5	2.4	150	NB	NB	NB
ATP binding peptide	1	6.8	17	40	190	>5000	>5000	>5000	>5000

NB: no binding

Fig. 5. Comparisons of binding specificities of an RNA aptamer, a *md*RNA aptamer, and a peptide that bind to adenosine-5'-triphosphate (ATP).

Similarly, a modified DNA aptamer that can enantioselectively recognize an R-isomer of a thalidomide analog, loses its binding activity ($K_d = 1 \times 10^{-6}$ M) if aminohexyl-carbamoylmethyl groups, introduced as a foreign functionality, are removed; *i.e.*, the modified uracil bases are replaced with natural thymine bases [Shoji *et al.*, 2007]. Interestingly, despite the high enantioselectivity of this aptamer, the binding affinity did not decrease if the hydrogen group at the asymmetric carbon was replaced with a methyl group. According to a model that simulated a stable conformation and that was confirmed by experimental results, the hydrogen group projected outward into the water solution. The thalidomide analog is stacked with an adenine base, and the foreign amide group forms a hydrogen-bonding network together with the thalidomide analog and neighboring base, which provides stability to the complex. Thus, the chirality of such a highly symmetric low-molecular-weight compound could clearly be recognized with the assistance of the modified groups that were introduced.

5. Development of advanced screening system

Although the SELEX method has been successfully used to create nucleic acid aptamers, further improvements are required to make the screening process much more rapid and convenient. Nucleic acid aptamers that show high activities *in vitro* are not always as effective *in vivo*. Alternative evolutional approaches to create nucleic acid aptamers that can function in living cells and organisms are also being developed.

5.1 Capillary electrophoresis-SELEX

Capillary electrophoresis (CE)-SELEX and associated improved methods have been developed to obtain protein binding nucleic acid aptamers quickly and easily (Fig. 6a) [Mendonsa & Bowser, 2005]. The common feature of these methods is the use of CE in the affinity separation process, whereas solid supports, such as sepharose gel and nitrocellulose membranes, are used for conventional SELEX methods. Association and dissociation reactions between the target and nucleic acids occur in solution; therefore, enrichment of undesired sequences that bind to the solid support would be avoidable. In addition, use of an automated CE device with high resolving power can greatly reduce the number of repeated cycles of the screening operation. Conventional SELEX typically requires 8–15 selection cycles. CE-SELEX can complete the process in only 2–4 rounds. To date, various DNA aptamers that bind to the human immunodeficiency virus transferase (HIV-RT), immunoglobulin E (IgE), mismatch repair protein (MutS), and ricin, among others, have been obtained by CE-SELEX methods. These resulting DNA aptamers possess a sufficient binding ability as a specific binder. Their K_d values are 0.18 nM, 23 nM, 3.6 nM, and 58 nM, respectively [Mosing *et al.*, 2005; Mendonsa & Bowser, 2004; Drabovich *et al.*, 2005; Tang *et al.*, 2006;].

For CE-SELEX, an uncoated fused-silica capillary can be used because both nucleic acid and the inner wall of the capillary are negatively charged during electrophoresis. Capillaries in which the inner wall is coated are often used for CE separation analyses of biomolecules to suppress peak tailing due to interaction between analytes and the inner wall. The length of the oligonucleotide for the initial library is about 30 to 40 mer including 70 to 90 mer of the random sequence region. In many cases, the 5'-end of the oligonucleotide is labeled with a fluorophore, such as fluorescein (FAM), to detect the active species in the library using a laser induced fluorescence (LIF) detector, which is far more sensitive than a UV-absorbance detector.

In a typical CE-SELEX method of protein binding DNA aptamers, the library is incubated with the target. The mixture is injected into a capillary and subsequently separated using nonequilibrium CE of equilibrium mixtures (NECEEM). The sample and detector are set on the cathode and anode side, respectively, because electroosmotic flow (EOF) toward anode from cathode is generated when voltage is applied (Fig. 6b). Small and positively charged substances would migrate faster in the capillary. In many cases, DNA that forms a complex with the target protein would reach the detector prior to unbound free DNA, resulting in a clear separation between active and inactive species. However, if the molecular weight of the target protein is considerably large, at times, peaks of bound DNA and unbound DNA get closer and cannot be separated enough to conduct the screening. In such cases, the capillary with which the inner wall is coated, *e.g.*, polyvinyl alcohol is used for the separation. Analytes migrate toward the cathode from the anode in the running buffer at a pH higher than the isoelectric point (pI) of the target protein because EOF does not generate in the coated capillary. Therefore, contrary to the case using a fused-silica capillary, the sample and detector are set on the anode side and the cathode side, respectively.

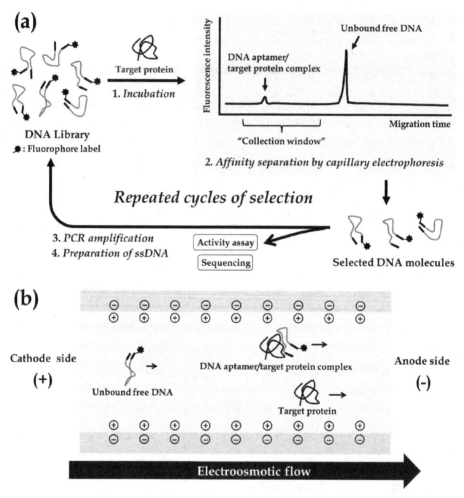

Fig. 6. General scheme of the CE-SELEX method (a); Affinity separation of the bund DNA from the unbound DNA using free zone CE (b).

During electrophoresis, association and dissociation between DNA and target protein is in a non-equilibrium state, and DNA gradually dissociates from the target protein. Therefore, DNA aptamers with a slow dissociation rate are expected to be obtained if a longer capillary is used. However, the migration time is proportional to the length of the capillary. The injection volume of the sample could be increased if a capillary with a larger inner diameter (i.d.) is used. However, the use of a thick capillary makes the release of Joule heating inefficient. Joule heating should be avoided because the active structure of the DNA aptamer would be destabilized due to heat denaturation. Thus, optimizations of capillary length and i.d. are important to obtain the target DNA aptamer efficiently. In many cases, capillaries with a length of approximately 31–80 cm and an i.d. of 50 or 75 μm are used.

Limitation of the library size is a defect of CE-SELEX; only several nanoliters of sample mixture can be injected per affinity separation. In the conventional SELEX method, the initial library normally contains miscellaneous sequences with 10^{13-15} diversities. If CE-SELEX were performed using a library of the same size, a sample mixture with a concentration of several tens to several hundreds mM has to be injected. With such a high DNA concentration, it is very difficult to separate the DNA aptamer/target protein complex from the free DNA by CE. Therefore, for CE-SELEX, the size of initial library is approximately 10^{10-13}.

Regardless of this limitation, CE-SELEX provides an effective methodology to obtain target DNA aptamers with a high incidence rate in a few selection rounds and there is no requirement of negative selection. For example, in the screening of IgE binding DNA aptamer by CE-SELEX, most sequences obtained in the enriched pool showed binding activity; almost 100% enrichment was achieved. Such fast enrichments of active species have not been reported for the conventional SELEX method in the literature; the enrichment has rarely been found to be above 50% within the first few rounds. Furthermore, non-SELEX selection, with no need to repeat tiresome selection cycles, has been developed as an advanced type of CE-SELEX. Thus, the generation rate for nucleic acid aptamers has dramatically increased, and the process has become more convenient by applying CE, with high resolving power, to the SELEX method.

5.2 Cell-SELEX

Cell-SELEX is a methodology used to obtain nucleic acid aptamers that specifically bind to a particular cell line [Daniels et al., 2003]. This method does not require prior knowledge of targets and would, in principle, generate nucleic acid aptamers for multiple targets simultaneously on the cell surface. Until now, nucleic acid aptamers specific to a number of cell lines, including a T-cell line (human acute lymphoblastic leukemia), B-cell line (human Burkitt lymphoma), and mouse liver hepatoma cell line, have been reported [Shangguan et al., 2006; Raddatz et al., 2008; Shangguan et al., 2008]. One problem with the conventional SELEX technique is that undesired sequences bind to the solid support such that the target becomes immobilized, causing negative selection to be performed. In cell-SELEX, sequences bound to the common cell matrix also need to be excluded. Therefore, in many cases, counter selection using a cell line akin to the target cell line is performed before or after selection using the target.

In the screening of cell-specific aptamers, an ssDNA library containing miscellaneous sequences with 10^{14-15} diversities is incubated with the target cell. Unbound sequences are subsequently removed by washing or centrifugation and bound sequences are amplified by polymerase chain reaction followed by preparation of ssDNA for the next selection round. The cell-specific aptamers can be obtained by repeating these processes approximately 20 times. For the present selection, living cells were used, although the sample was partially contaminated with dead cells. Dead cells are known to strongly and independently absorb nucleic acid sequences, interfering with enrichment of the target aptamers. To solve this problem, a method using fluorescence-activated cell sorting (FACS) has been developed [Raddatz *et al.*, 2008]. Use of FACS enables dead cells to be removed and active and inactive species to be separated at the same time.

There are many proteins that cause gene expression and a cellular response through signal transductions on the cell surface. Using nucleic acid aptamers specific to these proteins as targets, development of inhibitors that suppress their actions to achieve cell-specific drug delivery has been attempted. To date, nucleic acid aptamers that specifically bind to the cytokine interleukin-17 receptor and human RET receptor tyrosine kinase, for example, have been created by cell-SELEX, and their specific inhibition of the target proteins has been confirmed experimentally [Chen *et al.*, 2011]. As for examples of drug-delivery systems, conjugations of cell membrane protein binding nucleic acid aptamers to siRNAs, toxic proteins, nanoparticles, and antitumor drugs have been developed [McNamara *et al.*, 2006; Chu *et al.*, 2006]. Among these, modified nucleic acids, *e.g.*, RNAs with 2'-fluoro and 2'-amino nucleotides and phosphorothioate DNAs, which possess enhanced biostability, are included.

In the conventional SELEX technique using purified proteins as targets, it is anticipated that the nucleic acid aptamer obtained may not be able to bind to the target on the living cell. In contrast, the cell-SELEX can circumvent this problem because screening is performed using actual living cells. By choosing an appropriate cell for the counter selection, researchers can select nucleic acid aptamers targeting either known or unknown molecules, as well as the matrix structure particular to a certain cell line. Thus, cell-SELEX has provided a viable means to create nucleic acid aptamers with significant potential for *in vivo* application.

6. Outlook

The application range of nucleic acid aptamers can be very broad because of their notable advantages over antibodies. For example, nucleic acid aptamers can be obtained by random screening without using laboratory animals, and they can be produced by organic synthesis at low cost. While further technical innovations are still required to bring the performances of nucleic acid aptamers close to those of antibodies, many applications in bioanalyses ranging from test tube assays such as colorimetric sensing to imaging of living cells and tissues as well as therapeutic drugs have been investigated because of their aforementioned advantages. For *in vitro* assays, natural nucleic acid aptamers are used in many cases; however, chemical modifications assume increased importance in *in cell* and *in vivo* analyses to avoid degradation by nucleases. Therefore, development of rapid/facile screening techniques and *in vivo* screening techniques for creating modified nucleic acid aptamers will contribute to further advances in current bioanalytical measurements. In omics studies for non-nucleic acid molecules such as proteins, peptides, sugars, metabolites, until date, methodologies based on mass spectrometry are mainstreams. However, DNA molecules can readily be amplified by polymerase reactions such as PCR and rolling circle amplification, which generate 10^6–10^9 copies of an original sequence. As observed in a riboswitch, RNA molecules cause dynamic structural conversion by association of such small molecules like metabolites to provide multiple transcripts as a molecular offspring. Technologies involving 10^8 or more parallel sequencing and single molecule sequencing reactions are currently established. Considering these unique features of nucleic acid molecules and recent progress of sequencing technologies, aptamer/sequencing based analyses, which enable integrated omics analyses at single-cell level, may be developed, and these may replace the present methods for omics analyses in future.

7. Acknowledgment

This work was supported in part by a Grant for Industrial Technology Research from the New Energy and Industrial Technology Development Organization (NEDO) of Japan; Grants-in-Aid for Scientific Research, the "Core Research" project (2009–2014), and the "Academic Frontier" project (2004–2009) from the Ministry of Education, Culture, Sports, Science and Technology, Japan; the Hirao Taro Foundation of the Konan University Association for Academic Research; and the Long-Range Research Initiative (LRI) Project of the Japan Chemical Industry Association.

8. References

Andreola, M. L., Calmels, C., Michel, J., Toulmé, J. J. & Litvak, S. (2000). Towards the selection of phosphorothioate aptamers optimizing *in vitro* selection steps with phosphorothioate nucleotides, *Eur. J. Biochem.* 267(16), 5032–5040.

Aurup, H., Williams, D. M. & Eckstein F. (1992). 2'-Fluoro- and 2'-amino-2'-deoxynucleoside 5'-triphosphates as substrates for T7 RNA polymerase, *Biochemistry* 31(40): 9636–9641.

Bagalkot, V., Farokhzad, O. C., Langer, R. & Jon, S. (2006). An aptamer-doxorubicin physical conjugate as a novel targeted drug-delivery platform, *Angew. Chem. Int. Ed.*, 45(48): 8149–8152.

Bagalkot, V., Zhang, L., Levy-Nissenbaum, E., Jon, S., Kantoff, P. W., Langer, R. & Farokhzad, O. C. (2007). Quantum dot-aptamer conjugates for synchronous cancer imaging, therapy, and sensing of drug delivery based on bi-fluorescence resonance energy transfer, *Nano Lett.*, 7(10): 3065–3070.

Battersby, T. R., Ang, D. N., Burgstaller, P., Jurczyk, S. C., Bowser, M. T., Buchanan, D. D., Kennedy, R. T. & Benner, S. A. (1999). Quantitative analysis of receptors for adenosine nucleotides obtained *via in vitro* selection from a library incorporating a cationic nucleotide analog, *J. Am. Chem. Soc.*, 121(42): 9781–9789.

Berezovski, M., Musheev, M., Drabovich, A. & Krylov, S. N. (2006). Non-SELEX selection of aptamers, *J. Am. Chem. Soc.*, 128(5): 1410–1411.

Breaker, R. R. & Joyce, G. F. (1994). A DNA enzyme that cleaves RNA, *Chem Biol.* 1(4): 223–229.

Bock, L. C., Griffin, L. C., Latham, J. A., Vermaas, E. H. & Toole, J. J. (1992). Selection of single-stranded DNA molecules that bind and inhibit human thrombin, *Nature* 355(6360): 564–566.

Chaput, J. C. & Szostak, J. W. (2004). Evolutionary optimization of a nonbiological ATP binding protein for improved folding stability, *Chem Biol.*, 11(6): 865–874.

Chen, L., Li, D. Q., Zhong, J., Wu, X.L., Chen, Q., Peng, H. & Liu, S. Q. (2011). IL-17RA aptamer-mediated repression of IL-6 inhibits synovium inflammation in a murine model of osteoarthritis, *Osteoarthr. Cartil.*, 19(6): 711–718.

Chu, T.C., Marks, J.W. III, Lavery, L. A., Faulkner, S., Rosenblum, M. G., Ellington, A. D. & Levy, M. (2006). Aptamer:toxin conjugates that specifically target prostate tumor cells, *Cancer Res.*, 66(12): 5989–5992.

Daniels, D. A., Chen, H., Hicke, B. J., Swiderek, K. M., Gold, L. (2003). A tenascin-C aptamer identified by tumor cell SELEX: systematic evolution of ligands by exponential enrichment, *Proc. Natl. Acad. Sci. U S A*, 100(26): 15416–15421.

Drabovich, A., Berezovski, M. & Krylov, S. N. (2005). Selection of smart aptamers by equilibrium capillary electrophoresis of equilibrium mixtures (ECEEM), *J. Am. Chem. Soc.*, 127(32): 11224–11225.

Ellington, A. E. & Szostak, J. W. (1990). *In vitro* Selection of RNA molecules that bind specific ligands, *Nature* 346(6287): 818–822.

Geysen, H. M., Schoenen, F., Wagner, D. & Wagner, R. (2003). Combinatorial compound libraries for drug discovery: an ongoing challenge, *Nat. Rev. Drug Discov.* 2(3): 222–230.

Guerrier-Takada, C., Gardiner, K., Marsh, T., Pace, N. & Altman, S. (1983). The RNA moiety of ribonuclease P is the catalytic subunit of the enzyme, *Cell* 35(3): 849–857.

Hollenstein, M., Hipolito, C. J., Lam, C. H. & Perrin, D. M. (2009). A self-cleaving DNA enzyme modified with amines, guanidines and imidazoles operates independently of divalent metal cations (M^{2+}). *Nucleic Acids Res.* 37(5): 1638–1649.

Holmberg, A., Blomstergren, A, Nord, O., Lukacs, J. Lundeberg, J. & Uhlén, M. (2005). The biotin-streptavidin interaction can be reversibly broken using water at elevated temperatures, *Electrophoresis*, 26(3): 501–510.

Huizenga, D. E., & Szostak, J. W. (1995). A DNA aptamer that binds adenosine and ATP, *Biochemistry*, 34(2): 656–665.

Jensen, K. B., Atkinson, B. L., Willis, M. C., Koch, T. H. & Gold, L. (1995). Using *in vitro* selection to direct the covalent attachment of HIV-1 Rev protein to high affinity RNA ligands. *Proc. Natl. Acad. Sci. USA* 92(26): 12220–12224.

Jhaveri, S., Olwin, B. & Ellington, A.D. (1998). *In vitro* selection of phosphorothiolated aptamers. *Bioorg. Med. Chem. Lett.* 8(17): 2285–2290.

Kasahara, Y., Kitadume, S., Morihiro, K., Kuwahara, M., Ozaki, H., Sawai, H., Imanishi T. & Obika S. (2010). Effect of 3'-end capping of aptamer with various 2',4'-bridged nucleotides: Enzymatic post-modification toward a practical use of polyclonal aptamers, *Bioorg. Med. Chem. Lett.* 20(5): 1626-1629.

Klussmann, S., Nolte, A., Bald, R., Erdmann, V. A. & Furste, J. P. (1996). Mirror-image RNA that binds ᴅ-adenosine, *Nat. Biotechnol.* 14(9): 1112–1115.

Kubik, M. F., Stephens, A. W., Schneider, D., Marlar, R. A. & Tasset, D. (1994). High-affinity RNA ligands to human α-thrombin. *Nucleic Acids Res.*, 22(13): 2619–2626.

Kujau, M. J. & Wölfl, S. (1998). Intramolecular derivatization of 2'-amino-pyrimidine modified RNA with functional groups that is compatible with re-amplification, *Nucleic Acids Res.* 26(7): 1851–1853.

Kuwahara, M., Nagashima, J., Hasegawa, M., Tamura, T., Kitagata, R., Hanawa, K., Hososhima, S., Kasamatsu, T., Ozaki, H. & Sawai, H. (2006). Systematic characterization of 2'-deoxynucleoside-5'-triphosphate analogs as substrates for DNA polymerases by polymerase chain reaction and kinetic studies on enzymatic production of modified DNA, *Nucleic Acids Res.* 34(19): 5383–5394.

Kuwahara, M., Obika, S., Nagashima, J., Ohta, Y., Suto, Y., Ozaki, H., Sawai, H. & Imanishi, T. (2008). Systematic analysis of enzymatic DNA polymerization using oligo-DNA templates and triphosphate analogs involving 2',4'-bridged nucleosides, *Nucleic Acids Res.* 36(13): 4257–4265.

Kuwahara, M., Takeshima, H., Nagashima, J., Minezaki, S., Ozaki, H. & Sawai, H. (2009). Transcription and reverse transcription of artificial nucleic acids involving backbone modification by template-directed DNA polymerase reactions, *Bioorg. Med. Chem.* 17(11): 3782–3788.

Lato, S. M., Ozerova, N. D. S., He, K., Sergueeva, Z., Shaw, B. R. & Burke, D. H. (2002). Boron-containing aptamers to ATP. *Nucleic Acids Res.* 30(6): 1401–1407.

Lee, S. E., Sidorov, A., Gourlain, H., Mignet, N., Thorpe, S. J., Brazier, J. A., Dickman, M. J., Hornby, D. P., Grasby, J. A. & Williams, D. M. (2001). Enhancing the catalytic repertoire of nucleic acids: a systematic study of linker length and rigidity, *Nucleic Acids Res.* 29(7): 1565–1573.

Lin, Y., Qiu, Q., Gill, S. C. & Jayasena, S. D. (1994). Modified RNA sequence pools for *in vitro* selection. *Nucleic Acids Res.* 22(24): 5229–5234.

Mandal, M., Boese, B., Barrick, J. E., Winkler, W. C. & Breaker, R. R. (2003). Riboswitches control fundamental biochemical pathways in Bacillus subtilis and other bacteria, *Cell*, 113(5): 577–586.

McNamara, J. O. II, Andrechek, E. R., Wang, Y., Viles, K. D., Rempel, R. E., Gilboa, E., Sullenger, B. A & Giangrande, P. H. (2006). Cell type-specific delivery of siRNAs with aptamer-siRNA chimeras, *Nat Biotechnol.*, 24(8): 1005–1015.

Mendonsa, S.D. & Bowser, M. T. (2004). *In vitro* selection of high-affinity DNA ligands for human IgE using capillary electrophoresis, *Anal. Chem.*, 76(18): 5387–5392.

Mendonsa, S. D. & Bowser, M. T. (2005). *In vitro* selection of aptamers with affinity for neuropeptide Y using capillary electrophoresis, *J. Am. Chem. Soc.*, 127(26): 9382–9383.

Mosing, R. K., Mendonsa, S. D. & Bowser, M. T. (2005). Capillary electrophoresis-SELEX selection of aptamers with affinity for HIV-1 reverse transcriptase, *Anal. Chem.*, 77(19): 6107–6112.

Nolte, A., Klussmann, S., Bald, R., Erdmann, V.A. & Fürste, J. P. (1996). Mirror-design of L-oligonucleotide ligands binding to L-arginine, *Nat. Biotechnol.* 14(9): 1116–1119.

Pagratis, N. C., Bell, C., Chang, Y. F., Jennings, S., Fitzwater, T., Jellinek, D. & Dang, C. (1997). Potent 2'-amino-, and 2'-fluoro-2'-deoxyribonucleotide RNA inhibitors of keratinocyte growth factor. *Nat. Biotechnol*, 15(1): 68–73.

Perrin, D. M., Garestier, T. & Hélène, C. (2001). Bridging the gap between proteins and nucleic acids: a metal-independent RNAse A mimic with two protein-like functionalities, *J. Am. Chem. Soc.* 123(8): 1556–1563.

Raddatz, M. S., Dolf, A., Endl, E., Knolle, P., Famulok, M. & Mayer, G. (2008). Enrichment of cell-targeting and population-specific aptamers by fluorescence-activated cell sorting, *Angew. Chem. Int. Ed.*, 47(28): 5190–5193.

Robertson, D. L. & Joyce, G. F. (1990). Selection *in vitro* of an RNA enzyme that specifically cleaves single-stranded DNA, *Nature* 344(6265): 467–468.

Ruckman, J., Green, L. S., Beeson, J., Waugh, S., Gillette, W. L., Henninger, D. D., Claesson-Welsh, L. & Janjić, N. (1998). 2'-Fluoropyrimidine RNA-based aptamers to the 165-amino acid form of vascular endothelial growth factor (VEGF165). Inhibition of receptor binding and VEGF-induced vascular permeability through interactions requiring the exon 7-encoded domain, *J. Biol. Chem.* 273(32): 20556–20567.

Sakthivel, K. & Barbas, C. F. III. (1998). Expanding the potential of DNA for binding and catalysis: highly functionalized dUTP derivatives that are substrates for thermostable DNA polymerases, *Angew. Chem. Int. Ed.* 37(20): 2872–2875.

Sazani, P. L., Larralde, R. & Szostak, J. W. (2004). A small aptamer with strong and specific recognition of the triphosphate of ATP, *J. Am. Chem. Soc.*, 126(27): 8370–8371.

Shangguan, D., Li, Y., Tang, Z., Cao, Z. C., Chen, H. W., Mallikaratchy, P., Sefah, K., Yang, C. J., Tan, W. (2006). Aptamers evolved from live cells as effective molecular probes for cancer study, *Proc. Natl. Acad. Sci. U S A*, 103(32): 11838–11843.

Shangguan, D., Meng, L., Cao, Z. C., Xiao, Z., Fang, X., Li, Y., Cardona, D., Witek, R. P., Liu, C. & Tan, W. (2008). Identification of liver cancer-specific aptamers using whole live cells, *Anal. Chem.*, 80(3): 721–728.

Shoji, A., Kuwahara, M., Ozaki, H. & Sawai, H. (2007). Modified DNA aptamer that binds the (R)-isomer of a thalidomide derivative with high enantioselectivity, *J. Am. Chem. Soc.*, 129(5): 1456–1464.

Sidorov, A. V., Grasby, J. A. & Williams, D. M. (2004). Sequence-specific cleavage of RNA in the absence of divalent metal ions by a DNAzyme incorporating imidazolyl and amino functionalities. *Nucleic Acids Res.* 32(4): 1591–1601.

Smith, E. A., Thomas, W. D., Kiessling, L. L. & Corn, R. M. (2003). Surface plasmon resonance imaging studies of protein-carbohydrate interactions, *J. Am. Chem. Soc.*, 125(20): 6140–6148.

Smith, G.P & Petrenko, V. A. (1997). Phage display, *Chem. Rev.* 97(2): 391–410.

Tasara, T., Angerer, B., Damond, M., Winter, H., Dorhofer, S., Hubscher, U. & Amacker, M. (2003). Incorporation of reporter molecule-labeled nucleotides by DNA polymerases. II. High-density labeling of natural DNA, *Nucleic Acids Res.* 31(10): 2636–2646.

Tang, J., Xie, J., Shao, N. & Yan, Y. (2006). The DNA aptamers that specifically recognize ricin toxin are selected by two in vitro selection methods, *Electrophoresis*, 27(7): 1303–1311.

Tetin, S. Y., Swift, K. N. & Matayoshi, E. D. (2002). Measuring antibody affinity and performing immunoassay at the single molecule level, *Anal. Biochem.*, 307(7): 84–91.

Tuerk, C. & Gold, L. (1990). Systematic evolution of ligands by exponential enrichment: RNA ligands to bacteriophage T4 DNA polymerase., *Science* 249(4968): 505–510.

Uphoff, K. W., Bell, S. D. & Ellington, A. D. (1996). *In vitro* selection of aptamers: the dearth of pure reason, *Curr. Opin. Struct. Biol.* 6(3): 281–288.

Vaish, N. K., Larralde, R., Fraley, A. W., Szostak, J. W. & McLaughlin, L. W. (2003). A novel, modification-dependent ATP-binding aptamer selected from an RNA library incorporating a cationic functionality, *Biochemistry*, 42(29): 8842–8851.

Williams, K. P. & Liu, X. -H. (1997). Schumacher, T. N. M., Lin, H. Y., Ausiello, D. A., Kim, P. S. & Bartel., D. P. (1997). Bioactive and nuclease-resistant ʟ-DNA ligand of vasopressin. *Proc. Natl. Acad. Sci. USA*, 94(21): 11285–11290.

Wilson, C., Nix, J. & Szostak, J. (1998). Functional requirements for specific ligand recognition by a biotin-binding RNA pseudoknot, *Biochemistry*, 37(41): 14410–14419.

Winkler, W. C., Cohen-Chalamish, S. & Breaker, R. R. (2002). An mRNA structure that controls gene expression by binding FMN, *Proc. Natl. Acad. Sci. U S A.*, 99(25): 15908–15913.

Winkler, W. C., Nahvi, A. & Breaker, R. R. (2002). Thiamine derivatives bind messenger RNAs directly to regulate bacterial gene expression, *Nature* 419(6910): 952–956.

Winkler, W. C., Nahvi, A., Sudarsan, N., Barrick, J. E. & Breaker, R. R. (2003). An mRNA structure that controls gene expression by binding S-adenosylmethionine, *Nat. Struct. Biol.*, 10(9): 701–707.

Zaug, A. J. & Cech, T. R. (1986). The intervening sequence RNA of *Tetrahymena* is an enzyme, *Science* 231(4737): 470–475.

8

The Application of Pooled DNA Sequencing in Disease Association Study

Chang-Yun Lin[1] and Tao Wang[2]
[1]McDermott Center of Human Growth and Development and Department of Clinical Sciences, University of Texas Southwestern Medical Center, Dallas, TX,
[2]Department of Epidemiology and Population Health,
Albert Einstein College of Medicine, Bronx, NY,
USA

1. Introduction

Hundreds of common genetic variants related to the risk of human disease, such as diabetes, hypertension, bipolar, and Crohn's disease, have been successfully discovered by Genome-wide Association Studies (GWAS) (Barret et al., 2008; Hindorff, 2009; Thomas et al., 1991; WTCCC, 2007). Current GWAS are based on the strategy of linkage disequilibrium (LD) mapping, in which a sufficient number of single nucleotide polymorphism (SNP) markers are selectively genotyped to capture the genetic variation of the whole genome. However, there are two major issues related to the results of GWAS. First, the results only explain a small fraction of the heritability of complex diseases. One of the reasons may be that many functional variants, in particular rare variants, which are not directly genotyped in GWAS, have a weak LD with SNP markers, and hence are missed by GWAS (Iyengar et al., 2004; Manolio et al., 2009). Second, the identified associations in GWAS are often inconsistent between different populations. The reason for this may be the varied LD structures between markers and underlying causal variants among populations, resulting in associations can only be observed in specific populations.

To address these issues, an ideal approach is to directly sequence all the samples in a study (Bodmer & Bonilla, 2008). However, this is not a feasible option for the traditional sequencing technology, namely Sanger sequencing, which is extremely expensive and time consumption for sequencing thousands of samples required to achieve reasonable statistical power in a typical genetic association study.

Next generation sequencing (NGS) technology, also called parallel sequencing, is a revolutionary technology for biomedical research (Shendure & Ji, 2008). The production of large numbers of low-cost reads makes NGS useful for many applications. Today there are three commonly-used next-generation sequencing systems: namely Roche's (454) GS FLX Genome Analyzer marketed by Roche Applied Sciences, Illumina's Genome Analyzer" (GA), and Applied Biosystem's SOLiD system. Several new systems have either just been introduced or will become available soon (Metzker, 2010). One of the most important applications is to identify DNA variants, in particular rare variants, responsible for human diseases (Metzker, 2010). Now ten billion bases can be obtained routinely in a single run of NGS instrument and yields are expected to continually increase. The throughput of the smallest function unit, e.g., a single 'lane', can generate data amounting to many thousands fold coverage for a target region, which is far greater than what is needed for genotyping one individual as the individual genotype at a specific locus is expected to be accurately called at about 15-30 fold coverage. As such, it is feasible to simultaneously sequence targeted regions of multiple individuals with dramatic saving on cost and time.

To reduce the cost of large-scale association studies, one efficient approach is to sequence a large number of individuals together on a single sequence run. Two commonly-used approaches are available in disease association studies. Bar-coding ligates the DNA fragments of each sample to a short, sample-specific DNA sequence, and then sequences these DNA fragments from multiple subjects in one single sequencing run. In addition to allowing determining individual genotypes, bar-coding offers an additional advantage of reduction of sequencing variability (Craig et al. 2008). However, bar-coding at present has a limit of the multiplexing and the cost on the individual DNA amplification and sequencing template preparation could be substantial in large scale disease-association studies. Compared to bar-coding, simply pooling DNA samples is more cost-effective as it can fully make use of the high depth of sequencing and vastly reduce the efforts of sample preparation for thousands of individuals. Currently pooled DNA sequencing is particularly appealing due to its substantial cost and time-saving in large disease-association studies, i.e. pooled DNA sequencing (Shaw et al., 1998). With pooling, the sequencing throughput required per individual is much less than what is provided by a single run, and hence it is feasible to sequence multiple individuals together. For example, in a case-control study, the allele frequencies in a sample of 500 cases and 500 controls can be measured from two pooled samples, rather than from 1,000 individual samples, which represents an increase in efficiency of 500-fold.

Pooling was first used in genetic study in a case-control association study of HLA class II DR and DQ alleles in type I diabetes mellitus (Arnheim et al., 1985). Afterwards, it has been used for linkage studies in plants (Michelmore et al., 1991), for the homozygosity mapping of recessive diseases in inbred populations (Sheffield et al., 1994; Carmi et al., 1995; Nystuen et al., 1996; Scott et al., 1996), and for mutation detection (Amos et al., 2000). This strategy was also proposed for high-throughput SNP arrays (Ito et al., 2003; Shaw et al., 1998; Zeng & Lin, 2005) but it was not widely accepted as SNP array technology does not provide accurate estimates of the allele frequencies in the pooled samples. Recent next generation sequencing technology provides a high-throughput sequencing solution for examining functional variants directly. It might provide more accurate estimates of allele frequency, as shown by recent studies (Druley et al., 2009). Recently, one study adopted this strategy using 454 sequencing technology and identified associations of rare variants with insulin-dependent diabetes mellitus (Nejentsev et al., 2009). In a genome-wide analysis studies, two-stage design and DNA pooling could be used as a cost-efficient strategy to detect genetic variant regions (Chi et al., 2009; Skol et al., 2006; Wang et al., 2006; Zuo et al., 2006, 2008). In the first stage, a fraction of samples are genotyped for all SNPs and a case-control association test for each SNP is then conducted to select the most significant SNPs. In the second stage, the candidate SNPs from the first stage are further evaluated by genotyping. To reduce the cost of large-scale association studies in two-stage design, pools of DNA from many individuals have been successfully used in the first stage of the two-stage design (Bansal et al., 2002; Boss et al., 2009; Nejentsev et al., 2009; Norton et al., 2004; Sham et al., 2002). As suggested by Out et al. (2009), the use of a pooled DNA sample for targeted regions, NGS also can be an attractive cost- effective method to identify rare variants in candidate genes.

The data produced by next-generation sequencing is different from that of SNP-chips. Next-generation re-sequencing produces large amounts of short reads. After mapping to the reference genome, an alignment of reads across the targeted regions is obtained. A schematic example of re-sequencing data in case-control study is shown in Table 1. In this example, each case and control sample consists of two pools with two individuals in each pool. The two alleles (A and a) of each individual are shown in the "Genotype" column. Each allele appears a random number of times. Although NGS have the potential to discover the entire spectrum of sequence variations in a sample of well-phenotyped individuals, NGSs also present challenges. First, the error rate of these platforms is higher than conventional sequencing methods, and many errors are not random events (Johnson & Slatkin, 2008; Chaisson et al., 2009; Lynch, 2009; Bansal et al., 2010b). These errors may be frequent enough to obscure true associations or systematic enough to generate false-positive associations. Second, the data produced by next-generation sequencing often lose linkage disequilibrium (LD) information which is lost in pooled sequencing. As the result, the powerful analytic approaches that combine multiple rare variants to examine the disease association are not directly applicable to pooled sequencing, because these approaches require individual genotypes to account for the LD between SNPs. The current single locus analysis of pooled sequencing data could be very inefficient, in particular, for rare variants.

	Pool	Individual	Genotype	Read base
Case	Pool 1	1	A	A,A,A,a,A,A,A,A,A,A,A,A
			a	a,a,a,a,a,a,a,a,a,a,a,a
		2	A	A,A,A,A,A,A,A,A
			a	a,a,A,a,a,a,a,a,a,a,a
	Pool 2	3	a	a,a,a,a,a,a,a,a,a,a,a,a,a,a
			a	a,a,a,a,a,a,a,a,a,a,a,a,a
		4	A	A,A,A,A,A,A,A,A,A,A,A,A
			a	a,a,a,a,a,a,a,a,a,a,A,a,a
Control	Pool 3	5	a	a,a,a,a,a,a,a,a,a,a,a,
			a	a,a,a,a,a,a,a,a,a,a,a,a,a,a
		6	a	a,a,a,a,a,a,A,a,a,a,a,a,a,a,a
			a	a,a,a,a,a,a,a,a,a,a,a,a,a
	Pol 4	7	A	a,A,A,A,A,A,A,A,A,A,A,A
			a	a,a,a,a,A,a,a,a,a,a,a,a,a,a,a
		8	a	a,a,a,a,a,a,a,a,a,a,a,a,a,a
			a	a,a,a,a,a,a,a,a,a,a,a,a,a,a,a,a

Table 1. Example of re-sequencing data in case-control. Each case and control sample consists of two pools with two individuals in each pool.

In section 2 of this chapter, we will introduce some strategies of pooling design, including PI-deconvolution, shifted-transversal design, multiplexed scheme, and overlapping pools to recover LD information. Through these well-chosen pool designs, the variant carriers can be clearly identified, which greatly enhances the pooling efficiency. In section 3, we will introduce some statistical methods for the detection of variant and case-control association study to account for high-levels of sequencing errors. A briefly summary is added in the end of this chapter.

2. Strategies of pooling design

The main idea of pooling is to sequence DNA from several individuals together on a single run. Through the observed number of re-sequencing alleles, the allele frequency can be estimated. The simplest strategy is the naïve-pooling scheme, which is also called disjoint pooling. In naïve-pooling scheme, DNA was sequenced from several individuals on a single pool and each pool includes different individuals (Table 2). It offers insight into allele frequencies, but is not able to the identity of an allele carrier.

Recently, several strategies of well-chosen pools aiming to identify variant are proposed. In these designs, each individual is tested several times in different pools. This redundancy provides a potential increase in both sensitivity and specificity. We will introduce PI-deconvolution, shifted-transversal design, multiplexed scheme, and overlapping pools.

	Individuals															
	1	2	3	4	5	6	7	8	9	10	11	12	13	14	15	16
Pool 1	1	1	0	0	0	0	0	0	0	0	0	0	0	0	0	0
Pool 2	0	0	1	1	0	0	0	0	0	0	0	0	0	0	0	0
Pool 3	0	0	0	0	1	1	0	0	0	0	0	0	0	0	0	0
Pool 4	0	0	0	0	0	0	1	1	0	0	0	0	0	0	0	0
Pool 5	0	0	0	0	0	0	0	0	1	1	0	0	0	0	0	0
Pool 6	0	0	0	0	0	0	0	0	0	0	1	1	0	0	0	0
Pool 7	0	0	0	0	0	0	0	0	0	0	0	0	1	1	0	0
Pool 8	0	0	0	0	0	0	0	0	0	0	0	0	0	0	1	1

Table 2. Re-sequencing with naïve pooling scheme. A total of 16 individuals are divided into groups of two and pooled.

PI-deconvolution (Jin et al., 2006) The PI-deconvolution approach is a classic grid design. This strategy assigns individuals on an imaginary grid and construct pools by each row and each column. The individuals with variant then can usually be identified from the pattern of pools appearing variant. If there is a confounding among individuals, only a few candidates need to be retested. For example (Table 3), 16 individuals are arrayed on an imaginary grid and mixed in 8 pools, each containing 4 individuals (individuals 1, 2, 3, and 4 are in pool 1 and individuals 1, 5, 9, and 13 are in pool 5). If the pools 3 and 6 appeal variant, then individual 10 is the only variant carrier. If pools 2 and 7 also appear variant, we cannot distinguish whether the variant is from individuals 6 and 11 or from individuals 7 and 10. To resolve this confounding, we can add four additional pools, built along one of the grid's diagonals as indicated by the colors of the individuals. If the pink diagonal pool appears variant, individuals 6 and 11 are the variant carriers, whereas if both the orange and blue diagonal pools appear variant, the variant is from individuals 7 and 10. The author has validated the technique in three experimental contexts: protein chips, yeast two-hybrid assay, and drug resistance screening.

Shifted-Transversal Design (Thierry-Mieg, 2006) This method minimizes the co-occurrence of objects and constructs pools of constant-sized intersection. They proved that it allows unambiguous decoding of noisy experimental observations. It is highly flexible and can be tailored to function robustly in a wide range of experimental settings. Let $n \geq 2$, and consider the set $\mathcal{A}_n = \{A_0, \cdots, A_{n-1}\}$ of n Boolean variables. Let σ_q be the mapping of $\{0,1\}^q$ onto itself defined by:

	Pool 5	Pool 6	Pool 7	Pool 8
Pool 1	1	2	3	4
Pool 2	5	6	7	8
Pool 3	9	10	11	12
Pool 4	13	14	15	16

Table 3. Example for the strategy of PI-deconvolution. Samples are arranged in to an array, in which each row and each column represents a pool. In this example, 16 individuals are well-chosen into 8 pools.

$$\forall (x_1, \cdots, x_q) \in \{0,1\}^q, \sigma_q \begin{bmatrix} x_1 \\ x_2 \\ \vdots \\ x_q \end{bmatrix} = \begin{bmatrix} x_q \\ x_1 \\ \vdots \\ x_{q-1} \end{bmatrix}.$$

For every $j \in \{0, \cdots, q\}$, let M_j be a $q \times n$ Boolean matrix, defined by its columns $C_{j,0}, \cdots, C_{j,n-1}$ as follows:

$$C_{0,0} = \begin{bmatrix} 1 \\ 0 \\ \vdots \\ 0 \end{bmatrix}, and \ \forall i \in \{0, \cdots, n-1\} \ C_{j,i} = \sigma_q^{s(i,j)}(C_{0,0}),$$

where $s(i,j) = \sum_{c=0}^{\zeta} j^c \cdot \left\lfloor \frac{i}{q^c} \right\rfloor$ if $j < q$, and $s(i,q) = \left\lfloor \frac{i}{q^\zeta} \right\rfloor$, the semi-bracket denotes the integer part, and ζ denotes the smallest integer γ such that $q^{\gamma+1} \geq n$. Let $L(j)$ be the set of pools of which M_j is the matrix representation. For $k \in \{1, \cdots, q+1\}$, the transversal pooling design is defined as

$$STD(n; q; k) = \bigcup_{j=0}^{k-1} L(j).$$

For example, consider the variable $\mathcal{A}_9$ and $q = 3$, we have

$$M_0 = \begin{bmatrix} 1 0 0 1 0 0 1 0 0 \\ 0 1 0 0 1 0 0 1 0 \\ 0 0 1 0 0 1 0 0 1 \end{bmatrix}, M_1 = \begin{bmatrix} 1 0 0 0 0 1 0 1 0 \\ 0 1 0 1 0 0 0 0 1 \\ 0 0 1 0 1 0 1 0 0 \end{bmatrix}, M_2 = \begin{bmatrix} 1 0 0 0 1 0 0 0 1 \\ 0 1 0 0 0 1 1 0 0 \\ 0 0 1 1 0 0 0 1 0 \end{bmatrix},$$

$$M_3 = \begin{bmatrix} 1 1 1 0 0 0 0 0 0 \\ 0 0 0 1 1 1 0 0 0 \\ 0 0 0 0 0 0 1 1 1 \end{bmatrix}.$$

The corresponding layers of pools are the following:

Layer 0: $L(0) = \{\{A_0, A_3, A_6\}, \{A_1, A_4, A_7\}, \{A_2, A_5, A_8\}\}$,

Layer 1: $L(1) = \{\{A_0, A_5, A_7\}, \{A_1, A_3, A_8\}, \{A_2, A_4, A_6\}\}$,

Layer 2: $L(2) = \{\{A_0, A_4, A_8\}, \{A_1, A_5, A_6\}, \{A_2, A_3, A_7\}\}$,

Layer 3: $L(3) = \{\{A_0, A_1, A_2\}, \{A_3, A_4, A_5\}, \{A_6, A_7, A_8\}\}$.

The shifted-transversal design is

$$STD(9; 3; 2) = L(0) \cup L(1)$$
$$= \{\{A_0, A_3, A_6\}, \{A_1, A_4, A_7\}, \{A_2, A_5, A_8\}, \{A_0, A_5, A_7\}, \{A_1, A_3, A_8\}, \{A_2, A_4, A_6\}\}.$$

Suppose that a single variable in $\mathcal{A}_9$ is A_8. Then pools $\{A_2, A_5, A_8\}$ and $\{A_1, A_3, A_8\}$ are positive (appealing variant), which each proves that A_8 is positive.

Multiplexed scheme (Erlich et al., 2009) In this scheme, several pooling groups are created and the individuals are assigned to pools in each group by taking use of the Chinese remainder, one of the most ancient and fundamental in number theory (Andrews 1994; Ding et al. 1996; Cormen et al. 2001). To create a w pooling groups design, the rule of pooling for group k is

$$n_k = r_k \ (mod \ x_k),$$

which brings the $r_k, r_k + x_k, r_k + 2x_k, \cdots$ individuals to the rth pools of group k, where $0 < r_k \le x_k$ and $k = 1, \cdots, w$. According to the pooling pattern in each group, only a single, high-confidence solution would emerge for the vast majority of samples. For example (Table 4), let us create 2 groups for 20 individuals according to the following two pooling rules:

$$\begin{bmatrix} n_1 = r_1 (mod \ 5) \\ n_2 = r_2 (mod \ 8) \end{bmatrix}.$$

The total number of pools in this design is 13 (5+8). The corresponding pooling matrix is a 13×20 table and partitioned into two regions that correspond to the two pooling patterns. The staircase pattern (high- lighted in yellow) in each region is typically created by the multiplexed scheme. If pool 1 in group 1 and pool 6 in group 2 appear variant, then individual 6 can be identified as the variant carrier.

Overlapping pools (Prabhu & Pe'er, 2009) The central idea of overlapping pool design is that while sequencing DNA from several individual on a single pool, they also sequence DNA from a single individual on several pools. Individuals are assigned to pools in a manner so as to create a code: a unique set of pools for each individual. This set of pools on which an individual is sequenced defines a code word, or pool signature. If a variation is observed on the signature pools of one individual and on no other, then we identify the variant carrier. Based on the overlapping design, author proposed two algorithms for pool design: logarithmic signature designs and error-correcting designs. They showed that their designs guarantee high probability of unambiguous singleton carrier identification while maintaining the features of naïve pools in terms of sensitivity, specificity, and the ability to estimate allele frequencies.

Group	Pool	\multicolumn{20}{c}{Individual}

Group	Pool	1	2	3	4	5	6	7	8	9	10	11	12	13	14	15	16	17	18	19	20
1	1	1	0	0	0	0	1	0	0	0	0	1	0	0	0	0	1	0	0	0	0
	2	0	1	0	0	0	0	1	0	0	0	0	1	0	0	0	0	1	0	0	0
	3	0	0	1	0	0	0	0	1	0	0	0	0	1	0	0	0	0	1	0	0
	4	0	0	0	1	0	0	0	0	1	0	0	0	0	1	0	0	0	0	1	0
	5	0	0	0	0	1	0	0	0	0	1	0	0	0	0	1	0	0	0	0	1
2	1	1	0	0	0	0	0	0	0	1	0	0	0	0	0	0	1	0	0	0	0
	2	0	1	0	0	0	0	0	0	0	1	0	0	0	0	0	0	1	0	0	0
	3	0	0	1	0	0	0	0	0	0	0	1	0	0	0	0	0	0	1	0	0
	4	0	0	0	1	0	0	0	0	0	0	0	1	0	0	0	0	0	0	0	1
	5	0	0	0	0	1	0	0	0	0	0	0	0	1	0	0	0	0	0	0	0
	6	0	0	0	0	0	1	0	0	0	0	0	0	0	1	0	0	0	0	0	0
	7	0	0	0	0	0	0	1	0	0	0	0	0	0	0	1	0	0	0	0	0
	8	0	0	0	0	0	0	0	1	0	0	0	0	0	0	0	1	0	0	0	0

Table 4. The matrix is partitioned into two regions that correspond to the two pooling patterns. The staircase pattern (highlighted in yellow) is created by the multiplexed scheme.

	\multicolumn{16}{c}{Individual}

	1	2	3	4	5	6	7	8	9	10	11	12	13	14	15	16
Pool 1	0	0	0	0	0	0	0	0	1	1	1	1	1	1	1	1
Pool 2	0	0	0	0	1	1	1	1	0	0	0	0	1	1	1	1
Pool 3	0	0	1	1	0	0	1	1	0	0	1	1	0	0	1	1
Pool 4	0	1	0	1	0	1	0	1	0	1	0	1	0	1	0	1
Pool 5	1	1	1	1	1	1	1	1	0	0	0	0	0	0	0	0
Pool 6	1	1	1	1	0	0	0	0	1	1	1	1	0	0	0	0
Pool 7	1	1	0	0	1	1	0	0	1	1	0	0	1	1	0	0
Pool 8	1	0	1	0	1	0	1	0	1	0	1	0	1	0	1	0

Table 5. Sixteen distinct pool signatures are created using just eight pools in the over lapping design. If pools 1, 4, 6, and 7 appear variant, the variant carrier 10 can be identified.

3. Statistical methods for pooled DNS in GWAS

GWAS have successfully identified hundreds of variants that are associated with complex traits and pooled DNA sequencing has been considered a cost-effective approach for study rare variants in large populations. In this section, we discuss the statistical methods for the detection of variants and the case-control studies.

3.1 Detection of variants

SNPSeeker (Druley et al., 2009) This method (SNPSeeker) is an algorithm based on large deviation theory. It uses a seconder dependency error model for single-nucleotide polymorphism identification and takes into account the position in the sequencing read and the identity of the two upstream bases. This algorithm greatly improved the specificity of

SNP calling. The statistical models can be described as follows. Let $x \in \{A, C, G, T, N\}$ denote a observed base and $m \in \{A, C, G, T\}$ denote a base in the reference. The subset of nucleotides for each cycle j, sequencing run d and strand s can be defined as n i.i.d. random variables $x_{j,d,s,1}, x_{j,d,s,2}, \cdots, x_{j,d,s,n}$ and the empirical probability distribution can be written as

$$P_{j,d} = \left(\frac{As}{n}, \frac{Cs}{n}, \frac{Gs}{n}, \frac{Ts}{n}, \frac{Ns}{n}\right).$$

Under null hypothesis of no polymorphism at position i, the distribution of x is

$$Q_{j,d,s} = \sum_{m \in \{A,C,G,T\}} \Pr(x|M_i = m, j, d) * \Pr(M_i = m|s, \tau),$$

where $\Pr(x|M_i = m)$ is the probability of seeing a base x in the sequence at cycle position j on run d given that the original base at position i in the reference M_i is equal to m, and $\Pr(M_i = m|s, \tau)$ is the probability of observing nucleotide m in the reference sequence at position i, $M_i = m$ given the strand s and the true allele frequency vector τ. The cumulative p-value for each strand can be calculated by

$$\prod_d \prod_j 2^{-nD(P_{j,d,s}||Q_{j,d,s})}$$

where $D(P_{j,d}||Q_{j,d})$ is the Kullback-Leibler distance (Thomas & Joy, 1991) between $P_{j,d}$ and $Q_{j,d}$. Bonferoni-corrected is conducts for the total number of tests performed at each position in the reference sequence. The software for SNPseeker algorithm can be found at http://www.genetics.wustl.edu/rmlab/.

CRISP (Bansal, 2010a) This approach compares the distribution of allele counts across multiple pools using contingency tables and evaluates the probability of observing multiple non-reference base calls due to sequencing errors alone. The number of reads with the reference and alternate alleles at a particular position across the k pools can be modeled as a contingency table T^0 with two bases (rows) and k pools (columns) with row sums: $A = \sum_i a_i$ and $R - A = \sum_i r_i - a_i$ and column sums $r_i (1 \le i \le k)$:

	Pool 1	Pool 2	...	Pool k	Total
Reference base	$r_1 - a_1$	$r_2 - a_2$		$r_k - a_k$	$R - A$
Alternate base	a_1	a_2		a_k	A
Total	r_1	r_2		r_k	R

Under null hypothesis, the probability of the observed read can be defined as the probability of the table T^0:

$$P(T^0) = \binom{r_1}{a_1} \times \cdots \times \binom{r_k}{a_k} / \binom{R}{A}.$$

The p-value associated with the observed table T^0 is defined as the sum of all $2 \times k$ contingency tables with identical row and column sums that have equal or lower probability than the observed table:

$$p = \sum_{T \in \Gamma \text{ s.t. } P(T) \leq P(T^0)} P(T),$$

where Γ represents the set of all $2 \times k$ contingency tables with the same marginal sums as T^0. The p-value can be computed by Monte Carlo method:

1. Initialize $t = 0$
2. Initialize an array P of size R with $P[i] = p$ for $r_1 + \cdots + r_{p-1} \leq i \leq r_1 + \cdots + r_p$ $(1 \leq i \leq k)$
3. For $i = 1, \cdots, k$, set $a_i = 0$
4. For $i = 1, \cdots, N$, do the following:
   a. Set $P(T') = 1/\binom{R}{A}$
   b. For $a = 1, \cdots, A$, do
      i. Randomly select an integer r in the interval $[a, R]$
      ii. Set $j = P[r]$ and swap the elements $P[a]$ and $P[r]$
      iii. $P(T') = P(T') \times \frac{r_j - a_j}{a_j + 1}$, $a_j = a_j + 1$
   c. If $P(T') \leq P(T^0): t = t + 1$
   d. For any pool j chosen in step (2), set $a_j = 0$
5. The estimated p-value is $\frac{t+1}{N+1}$.

For example (Table 6), in contingency table (A), the p-value is 0.002 suggesting that the five base calls represent a rare SNP rather than sequencing errors; in contingency table (B), the p-value is 0.24 indicating that the presence of five alternate base calls in a single pool is likely due to sequencing error.

(A)	Pool 1	Pool 2	Pool 3	Pool 4
Reference base	42	40	44	50
Alternate base	0	5	0	0
(B)	Pool 1	Pool 2	Pool 3	Pool 4
Reference base	41	40	42	49
Alternate base	1	5	2	1

Table 6. Example of contingency tables for CRISP analysis.

3.2 Detect association base on case-control study

The model in case-control study can be described as follows. Let n_{ij} be the total number of chromosome segments of the region of interest in the jth pool of phenotypic group i, where $i = 1$ for case and 2 for control. Let z_{ij} be the unknown number of rare allele at a specific locus of interest. After re-sequencing, a total number of m_{ij} sequencing reads at the loci are observed and x_{ij} out of m_{ij} read report variant. We denote the random vector $Z_i = (Z_{i1}, \cdots, Z_{il_i})$ and $X_i = (X_{i1}, \cdots, X_{il_i})$. Let p_i be the frequency of the minor allele at this locus for group i. The question we are interested in the case-control study is whether this locus is associated with the disease. The hypothesis of the association can be written as:

$$H_0: p_1 = p_2 = p_0 \text{ versus } H_1: p_1 \neq p_1$$

Several statistical methods can be utilized for this test.

Fisher's exact test The allele frequencies of cases and controls are calculated from the observed numbers of total reads and from the number of reads reporting the variant; and the number of variants carried by each phenotypic group is estimated by $\hat{z}_i = \hat{p}_i n_i = (\sum x_{ij} / \sum m_{ij}) \times n_i$, where $\hat{p}_i$ is the estimated allele frequency for group i. The data are then summarized in a 2×2 table (Table 7) with the same row and column margins that have probabilities less than or equal to that of observed table. Although this test is simple to implement, it treats the estimated numbers of the rare variant as if they were observed without considering the uncertainty of such estimates. Thus, it may have an inflated Type I error rate. Second, the sampling scheme of Fisher's exact test is based on the hypergeometric distribution, which in principal requires both the column and row marginal totals of a 2×2 table are fixed, i.e., both the sample size (the number of cases and controls) and the number of variants and non-variants are fixed in Table I. The number of variants and non-variants are usually not fixed in a genetic case-control study. Fisher's exact test used in this way can become very conservative (Upton, 1982). Finally and most importantly, because next-generation sequencing has a relatively high rate of base-calling error and from sequence reads of pooled samples, it is difficult to distinguish true rare variants from such errors. For a rare variant whose frequency is not much higher than the error rate, the power to detect its association with a disease would be very low without adjusting for such error in the statistical method.

	Case	Controls	Total
Variant	$\hat{z}_1$	$\hat{z}_2$	$\hat{z}$
Non-variant	$n_1 - \hat{z}_1$	$n_2 - \hat{z}_2$	$n - \hat{z}$
Total	n_1	n_2	n

Table 7. A 2×2 table for Fisher's exact test.

Combined Z-test (Abraham et al., 2008) This method combines chi-square statistic and Z-statistic for testing the differences in mean allele frequencies between cases and controls. The general description of this statistic has been presented in (Sham et al., 2002; Macgregor, 2007; Kirov et al., 2009):

$$T_{comb} = \frac{\left(\bar{f}^{(2)} - \bar{f}^{(1)}\right)^2}{v_2 + v_1 + \varepsilon_2^2 + \varepsilon_1^2}$$

where $\bar{f}_i = \frac{1}{l_i}\sum_{j=1}^{l_i} f_j^{(i)}$ is the mean of the allele frequencies over K_i pool replicates, $v_i = \frac{\bar{f}_i(1-\bar{f}_i)}{2n_i}$ is the binomial sampling variance and n_i is number of controls and cases respectively ($i = 1, 2$) and $\varepsilon_i^2 = \frac{1}{n_i(n_i-1)}\sum_{j=1}^{l_j}\left(f_j^{(i)} - \bar{f}_i\right)^2$ is the square of the standard error due to experimental error. This method considers sampling error and experimental error, which is equivalent to a simplified version of the complex regression model suggested by Macgregor (2007).

Likelihood ratio test (Kim et al., 2010) To quantify he sequencing error rate in pooled sequencing, one approach is to include a control DNA sequence in each pool, which makes it possible to obtain the empirical distribution of the sequencing error of individual pools. Let $v \in \{A, C, G, T\}$ be the variant and $\gamma \in \{A, C, G, T\}$ be the reference allele. We define $\tau_{1,ij} = \Pr(v|\tau, i, j)$ to be the false positive error rate, i.e., the probability of reporting a variant given the reference base, and $\tau_{2,ij} = \Pr(v|v, i, j)$ to be the true-positive rate, i.e., the probability of reporting a variant given the variant based. Both the false-positive and true-positive rates can be estimated for each pool by the proportion of reads reporting the variant for the reference base and the variant base, respectively. The estimate of allele frequency with error can be calculated as

$$\theta_{ij} = \frac{z_{ij}}{n_{ij}}(1 - \tau_{2,ij}) + \left(1 - \frac{z_{ij}}{n_{ij}}\right)\tau_{1,ij}.$$

For next generation sequencing data, the likelihood can be computed as:

$$L(p|X) = \prod_{j=1}^{l_i} \sum_{z_{ij}=0}^{n_{ij}} Binominal(m_{ij}, x_{ij}, \theta_{ij})Binomial(n_{ij}, z_{ij}, p_i)$$

$$= \prod_{j=1}^{l_i} \sum_{z_{ij}=0}^{n_{ij}} \binom{m_{ij}}{x_{ij}} (\theta_{ij})^{x_{ij}}(1 - \theta_{ij})^{m_{ij}-x_{ij}} \binom{n_{ij}}{z_{ij}} p_i^{z_{ij}}(1 - p_i)^{n_{ij}-z_{ij}}$$

where θ_{ij} is as defined above. Then the Likelihood ratio statistic is computed as

$$LRT = -2log\frac{L(\hat{p}_0|X)}{L(\hat{p}_1, \hat{p}_2|X)}.$$

Reject H_0 if $LRT \geq \chi^2(1)$.

Differential test (Wang et al., 2010) Following the approaches of Liddell (1976) and Barry & Choongrak (1900), Wang et al. (2010) defined a test statistic by

$$T_{X_1,X_0} = \frac{\sum_j \frac{w_{1j}}{\sum w_{1j}}X_{ij}/m_{ij} - \sum_j \frac{w_{2j}}{\sum w_{2j}}X_{2j}/m_{2j}}{\sqrt{\hat{V}_X}}$$

where $\hat{V}_X = \sum_i \frac{\hat{p}_i}{\sum\left[\frac{m_{ij}n_{ij}}{m_{ij}+n_{ij}}\right]}$, $\hat{p}_i = \sum_j \left(\frac{w_{ij}}{\sum w_{ij}}\right)X_{ij}/m_{ij}$ for $i = 1,2$, and the weight w_{ij} is inversely proportional to the variance of X_{ij}/m_{ij}. Under null hypothesis, $\hat{p}_i = \hat{p}_0 = \sum_i \sum_j \left(\frac{w_{ij}}{\sum w_{ij}}\right)X_{ij}/m_{ij}$. The p-value can be calculated as

$$P\left(T_{X_1,X_2} \geq T_{x_1,x_2}|H_0\right)$$

$$= \sum_{X_{11}=0}^{m_{11}} \cdots \sum_{X_{1l_1}=0}^{m_{1l_1}} \sum_{X_{21}=0}^{m_{21}} \cdots \sum_{X_{2l_2}=0}^{m_{2l_2}} Pr\left(X_{11}, \cdots, X_{1l_1}|\hat{p}_1\right) Pr\left(X_{21}, \cdots, X_{2l_2}|\hat{p}_2\right)$$

$$\times I(|T_{X_1,X_2}| \geq |T_{x_1,x_2}|)$$

where

$$Pr\left(X_{i1}, \cdots, X_{il_i}|p_i\right) = \prod_{j=1}^{l_i} \sum_{z_{ij}=0}^{n_{ij}} \binom{m_{ij}}{x_{ij}} \left(\frac{z_{ij}}{n_{ij}}\right)^{x_{ij}} \left(1 - \frac{z_{ij}}{n_{ij}}\right)^{m_{ij}-x_{ij}} \binom{n_{ij}}{z_{ij}} p_i^{z_{ij}}(1 - p_{ij})^{n_{ij}-z_{ij}}$$

The two-side p-value for testing (1) is defined by the probability of observing an absolute value of the statistic T_{X_0,X_1} that is equal to or larger than the absolute value of the observed T_{X_1,X_0} under the null hypothesis.

If the sequencing error rates in pooled sequencing are considered, then

$$Pr(X_{i1}, \cdots, X_{il_i}|p_i) = \prod_{j=1}^{l_i} \sum_{z_{ij}=0}^{n_{ij}} \binom{m_{ij}}{x_{ij}} (\theta_{ij})^{x_{ij}} (1-\theta_{ij})^{m_{ij}-x_{ij}} \binom{n_{ij}}{z_{ij}} p_i^{z_{ij}} (1-p_i)^{n_{ij}-z_{ij}}$$

The parametric bootstrap (PB) procedure can be used to determine the p-value (Krishnamoorthy & Thomoson, 2004) by the following steps:

1. Estimating the sequencing error rate from the control sequence. $\hat{t}_{1,ij} = m_{v|\gamma}/(m_{v|\gamma} + m_{nv|\gamma})$, in which $m_{v|\gamma}$ and $m_{nv|\gamma}$ are the number of reads that report a variant or a non-variant for a reference base in the control sequence.
2. Estimate the allele frequency under null hypotheses by a weighted-average estimate across single pools with weight proportional to $n_{ij}m_{ij}/(n_{ij}+m_{ij})$ and the allele frequency of a single pool is estimated by $\hat{p}_{ij} = \sum_{z_{ij}=0}^{n_{ij}} \pi_{ij}z_{ij}/n_{ij}$ where

$$\pi_{ij} = \frac{\binom{m_{ij}}{x_{ij}} \left[\frac{z_{ij}}{n_{ij}}\hat{t}_{2,ij} + \frac{n_{ij}-z_{ij}}{n_{ij}}\hat{t}_{1,ij}\right]^{x_{ij}} \left[\frac{z_{ij}}{n_{ij}}(1-\hat{t}_{2,ij}) + \frac{n_{ij}-z_{ij}}{n_{ij}}(1-\hat{t}_{1,ij})\right]^{m_{ij}-x_{ij}}}{\sum_{z_{ij}=0}^{n_{ij}} \binom{m_{ij}}{x_{ij}} \left[\frac{z_{ij}}{n_{ij}}\hat{t}_{2,ij} + \frac{n_{ij}-z_{ij}}{n_{ij}}\hat{t}_{1,ij}\right]^{x_{ij}} \left[\frac{z_{ij}}{n_{ij}}(1-\hat{t}_{2,ij}) + \frac{n_{ij}-z_{ij}}{n_{ij}}(1-\hat{t}_{1,ij})\right]^{m_{ij}-x_{ij}}}$$

3. Calculating the test statistic

$$T_{X_1,X_0} = \frac{\sum_j \frac{w_{1j}}{\sum w_{1j}}X_{ij}}{m_{ij}} - \sum_j \frac{\frac{w_{2j}}{\sum w_{2j}}X_{2j}}{m_{2j}}}{\sqrt{\hat{V}_X}}$$

in which

$$w_{ij} = \frac{n_{ij}m_{ij}}{(\hat{t}_{2,ij} - \hat{t}_{1,ij})n_{ij}\hat{p}_0 + \hat{t}_{1,ij}n_{ij} + (\hat{t}_{2,ij} - \hat{t}_{1,ij})^2 m_{ij}\hat{p}_0}$$

and

$$\hat{V}_X = \sum_i \frac{1}{\sum(m_{ij}n_{ij})/[m_{ij}\hat{p}_0(\tau_{2,ij}-\tau_{1,ij})^2 + n_{ij}(\tau_{2,ij}\hat{p}_0 - \tau_{1,ij}\hat{p}_0 + \tau_{1,ij})]}$$

4. Sampling $\tilde{X}_i = (\tilde{x}_{i1}, \cdots)$ from $Pr(X_{i1}, \cdots, X_{il_i}|\hat{p}_0, \hat{t}_{i,ij}, \hat{t}_{2,ij})$ and calculating the statistic $T_{\tilde{X}_1,\tilde{X}_2}$.
5. Replicating (4) r times and estimating the p-value by the proportion of $|T_{\tilde{X}_1,\tilde{X}_2}| \geq |T_{X_1,X_2}|$

4. Summary

Pooled DNA has been widely used as a cost-effective strategy for genome wise association studies, which have successfully identified hundreds of variants that are associated with complex traits. Pooling scheme may provide less information comparing to the sequencing for each individual. However, though well-chosen pool designs as introduced in Section 2, there is still high chance to identify the variant carrier. Recently, next-generation sequencing technologies have made it feasible to sequence several human genomes entirely. SNPSeeker and CRISP are efficient statistical methods to detect the variant from the short read generated by NGS platforms. For case-control analysis, Fisher's exact test is common used but has been proved to be inappropriate. Several test methods such as Combined Z test, Likelihood ratio test, and differential test can be conducted for the association analysis.

5. References

Abraham R, Moskvina V, Sims R, Hollingworth P, Morgan A, Georgieva L, Dowzell K, Cichon S, Hillmer AM, O'Donovan MC, Williams J, Owen MJ, Kirov G. (2008). A genome-wide association study for late-onset Alzheimer's disease using DNA pooling. BMC Med Genomics. 29, 1-44.

Arnheim N, Strange C, Erlich H. (1985). Use of pooled DNA samples to detect linkage disequilibrium of polymorphic restriction fragments and human disease: studies of the HLA class II loci. 82, 6970-6974.

Amos CI, Frazier ML, Wang W. (2000). DNA pooling in mutation detection with reference to sequence analysis. Am J Hum Genet. 66, 1689-1692.

Bansal A, van den Boom D, Kammerer S, Honisch C, Adam G, Cantor CR, Kleyn P, Braun A. (2002). Association testing by DNA pooling: an effective initial screen. Proc Natl Acad Sci USA. 99, 16871-16874.

Bansal V. (2010a). A statistical method for the detection of variants from next-generation resequencing of DNA pools. Bioinformatics. 26, 318-324.

Bansal V, Harismendy O, Tewhey R, Murray SS, Schork NJ, Topol EJ, Frazer KA. (2010b). Accurate detection and genotyping of SNPs utilizing population sequencing data. Genome Res. 20, 537-545.

Barrett JC, Hansoul S, Nicolae DL, Cho JH, Duerr RH, Rioux JD, Brant SR, Silverberg MS, Taylor KD, Barmada MM, Bitton A, Dassopoulos T, Datta LW, Green T, Griffiths AM, Kistner EO, Murtha MT, Regueiro MD, Rotter JI, Schumm LP, Steinhart AH, Targan SR, Xavier RJ; NIDDK IBD Genetics Consortium, Libioulle C, Sandor C, Lathrop M, Belaiche J, Dewit O, Gut I, Heath S, Laukens D, Mni M, Rutgeerts P, Van Gossum A, Zelenika D, Franchimont D, Hugot JP, de Vos M, Vermeire S, Louis E; Belgian-French IBD Consortium; Wellcome Trust Case Control Consortium, Cardon LR, Anderson CA, Drummond H, Nimmo E, Ahmad T, Prescott NJ, Onnie CM, Fisher SA, Marchini J, Ghori J, Bumpstead S, Gwilliam R, Tremelling M, Deloukas P, Mansfield J, Jewell D, Satsangi J, Mathew CG, Parkes M, Georges M, Daly MJ. (2008). Genome-wide association defines more than 30 distinct susceptibility loci for Crohn's disease. Nat Genet. 40, 955-962.

Barry ES, Choongrak k. (1900). Exact Properties of Some Exact Test Statistics for Comparing Two Binomial Proportions . Journal of the American Statistical Association. 85, 146-155.

Bodmer W, Bonilla C. (2008). Common and rare variants in multifactorial susceptibility to common diseases. Nat Genet. 40, 695-701.

Boss Y, Bacot F, Montpetit A, Rung J, Qu HQ, Engert JC, Polychronakos C, Hudson TJ, Froguel P, Sladek R, Desrosiers M. (2009). Identification of susceptibility genes for complex diseases using pooling-based genome-wide association scans. Hum Genet. 125, 305-318.

Carmi R, Rokhlina T, Kwitek-Black AE, Elbedour K, Nishimura D, Stone EM, Sheffield VC. (1995). Use of a DNA pooling strategy to identify a human obesity syndrome locus on chromosome 15. Hum Mol Genet. 4, 9-13.

Chaisson MJ, Brinza D, Pevzner PA. (2009). De novo fragment assembly with short mate-paired reads: Does the read length matter? Genome Res. 19, 336-346.

Chi A, Schymick JC, Restagno G, Scholz SW, Lombardo F, Lai SL, Mora G, Fung HC, Britton A, Arepalli S, Gibbs JR, Nalls M, Berger S, Kwee LC, Oddone EZ, Ding J, Crews C, Rafferty I, Washecka N, Hernandez D, Ferrucci L, Bandinelli S, Guralnik J, Macciardi F, Torri F, Lupoli S, Chanock SJ, Thomas G, Hunter DJ, Gieger C, Wichmann HE, Calvo A, Mutani R, Battistini S, Giannini F, Caponnetto C, Mancardi GL, La Bella V, Valentino F, Monsurr MR, Tedeschi G, Marinou K, Sabatelli M, Conte A, Mandrioli J, Sola P, Salvi F, Bartolomei I, Siciliano G, Carlesi C, Orrell RW, Talbot K, Simmons Z, Connor J, Pioro EP, Dunkley T, Stephan DA, Kasperaviciute D, Fisher EM, Jabonka S, Sendtner M, Beck M, Bruijn L, Rothstein J, Schmidt S, Singleton A, Hardy J, Traynor BJ. (2009). A two-stage genome-wide association study of sporadic amyotrophic lateral sclerosis. Hum Mol Genet. 18, 1524-1532.

Craig DW, Pearson JV, Szelinger S, Sekar A, Redman M, Corneveaux JJ, Pawlowski TL, Laub T, Nunn G, Stephan DA, Homer N, Huentelman MJ. (2008). Identification of genetic variants using barcoded multiplexed sequencing. Nat Methods. 5, 887-893.

Druley TE, Vallania FL, Wegner DJ, Varley KE, Knowles OL, Bonds JA, Robison SW, Doniger SW, Hamvas A, Cole FS, Fay JC, Mitra RD. (2009). Quantification of rare allelic variants from pooled genomic DNA. Nat Methods. 6, 263-265.

Erlich Y, Chang K, Gordon A, Ronen R, Navon O, Rooks M, Hannon GJ. (2009). DNA Sudoku--harnessing high-throughput sequencing for multiplexed specimen analysis. Genome Res. 19, 1243-1253.

Hindorff LA, J. H. (2009). A Catalog of Published Genome-Wide Association Studies. National Human Genome Research Institute.

Ito T, Chiku S, Inoue E, Tomita M, Morisaki T, Morisaki H, Kamatani N. (2003). Estimation of haplotype frequencies, linkage-disequilibrium measures, and combination of haplotype copies in each pool by use of pooled DNA data. Am J Hum Genet. 72, 384-398.

Iyengar SK, Song D, Klein BE, Klein R, Schick JH, Humphrey J, Millard C, Liptak R, Russo K, Jun G, Lee KE, Fijal B, Elston RC. (2004). Dissection of genomewide-scan data in extended families reveals a major locus and oligogenic susceptibility for age-related macular degeneration. Am J Hum Genet. 74, 20-39.

Jin F, Hazbun T, Michaud GA, Salcius M, Predki PF, Fields S, Huang J. (2006). A pooling-deconvolution strategy for biological network elucidation. Nat Methods. 3, 183-189.

Johnson PL, Slatkin M. (2008). Accounting for bias from sequencing error in population genetic estimates. Mol Biol Evol. 25, 199-206.

Kim SY, Li Y, Guo Y, Li R, Holmkvist J, Hansen T, Pedersen O, Wang J, Nielsen R. (2010). Design of association studies with pooled or un-pooled next-generation sequencing data. Genet Epidemiol. 34, 479-491.

Kirov G, Zaharieva I, Georgieva L, Moskvina V, Nikolov I, Cichon S, Hillmer A, Toncheva D, Owen MJ, O'Donovan MC. (2009). A genome-wide association study in 574 schizophrenia trios using DNA pooling. Mol Psychiatry. 14, 796-803.

Krishnamoorthy K, Thomson J. (2004). A more powerful test for comparing two Poisson means. J Stat Plan Inference. 119, 23-35.

Liddell, D. (1976). Practical Tests of 2 × 2 Contingency Tables . Journal of the Royal Statistical Society. 25, 295-304.

Lynch M. (2009). Estimation of allele frequencies from high-coverage genome-sequencing projects. Genetics. 182, 295-301.

Macgregor S. (2007). Most pooling variation in array-based DNA pooling is attributable to array error rather than pool construction error. Eur J Hum Genet. 15, 501-504.

Maher, B. (2008). Peasonal genomes: the case of the missing heritability. 456, 18-21.

Manolio TA, Collins FS, Cox NJ, Goldstein DB, Hindorff LA, Hunter DJ, McCarthy MI, Ramos EM, Cardon LR, Chakravarti A, Cho JH, Guttmacher AE, Kong A, Kruglyak L, Mardis E, Rotimi CN, Slatkin M, Valle D, Whittemore AS, Boehnke M, Clark AG, Eichler EE, Gibson G, Haines JL, Mackay TF, McCarroll SA, Visscher PM. (2009). Finding the missing heritability of complex diseases. Nature. 461, 747-753.

Metzker ML. (2010). Sequencing technologies - the next generation. Nature reviews. 11, 31-46.

Michelmore RW, Paran I, Kesseli RV. (1991). Identification of markers linked to disease-resistance genes by bulked segregant analysis: a rapid method to detect markers in specific genomic regions by using segregating populations. Proc Natl Acad Sci U S A. 88, 9828-9832.

Nejentsev S, Walker N, Riches D, Egholm M, Todd JA. (2009). Rare variants of IFIH1, a gene implicated in antiviral responses, protect against type 1 diabetes. Science. 324, 387-389.

Norton N, Williams NM, O'Donovan MC, Owen MJ. (2004). DNA pooling as a tool for large-scale association studies in complex traits. Ann Med. 36, 146–152.

Nystuen A, Benke PJ, Merren J, Stone EM, Sheffield VC. (1996). A cerebellar ataxia locus identified by DNA pooling to search for linkage disequilibrium in an isolated population from the Cayman Islands. Hum Mol Genet. 5, 525-531.

Prabhu S, Pe'er I. (2009). Overlapping pools for high-throughput targeted resequencing. Genome Res. 19, 1254-1261.

Scott DA, Carmi R, Elbedour K, Yosefsberg S, Stone EM, Sheffield VC. (1996). An autosomal recessive non-syndromic- hearing-loss locus identified by DNA pooling using two inbred Bedouin kindreds. Am J Hum Genet. 59, 385-391.

Sham P, Bader JS, Craig I, O'Donovan M, Owen M. (2002). DNA Pooling: a tool for large-scale association studies. Nat Rev Genet. 3, 862-871.

Shaw SH, Carrasquillo MM, Kashuk C, Puffenberger EG, Chakravarti A. (1998). Allele frequency distributions in pooled DNA samples: applica- tions to mapping complex disease genes. . Genome Res. 8, 111-123.

Sheffield VC, Carmi R, Kwitek-Black A, Rokhlina T, Nishimura D, Duyk GM, Elbedour K, Sunden SL, Stone EM. (1994). Identification of a Bardet–Biedl syndrome locus on chromosome 3 and evaluation of an efficient approach to homozygosity mapping. Hum. Mol. Genet. . Hum. Mol. Genet. 3, 1331-1335.

Shendure J, Ji H. (2008). Next-generation DNA sequencing. Nat Biotechnol. 26, 1135-1145.

Skol AD, Scott LJ, Abecasis GR, Boehnke M. (2006). Joint analysis is more efficient than replication-based analysis for two-stage genome-wide association studies. Nat Genet. 38, 209–213.

Thierry-Mieg N. (2006). A new pooling strategy for high-throughput screening: the Shifted Transversal Design. BMC Bioinformatics. 19, 7-28.

Thomas G, Jacobs KB, Kraft P, Yeager M, Wacholder S, Cox DG, Hankinson SE, Hutchinson A, Wang Z, Yu K, Chatterjee N, Garcia-Closas M, Gonzalez-Bosquet J, Prokunina-Olsson L, Orr N, Willett WC, Colditz GA, Ziegler RG, Berg CD, Buys SS, McCarty CA, Feigelson HS, Calle EE, Thun MJ, Diver R, Prentice R, Jackson R, Kooperberg C, Chlebowski R, Lissowska J, Peplonska B, Brinton LA, Sigurdson A, Doody M, Bhatti P, Alexander BH, Buring J, Lee IM, Vatten LJ, Hveem K, Kumle M, Hayes RB, Tucker M, Gerhard DS, Fraumeni JF Jr, Hoover RN, Chanock SJ, Hunter DJ. (2009). A multistage genome-wide association study in breast cancer identifies two new risk alleles at 1p11.2 and 14q24.1 (RAD51L1). Nat Genet. 41, 579-584.

Thomas MC, Joy AT. (1991). Elements of Information Theory. Wily Interscience.

Upton, G. J. G. (1982). A comparison of alternative tests for the 2 x 2 comparative trial. J. R. Statist. Soc. A, 145, 86-105.

Wang H, Thomas DC, Pe'er I, Stram DO. (2006). Optimal two-stage genotyping designs for genome-wide association scans. Genet Epidemiol. 30, 356-368.

Wang T, Lin CY, Rohan TE, Ye K. (2010). Resequencing of pooled DNA for detecting disease associations with rare variants. Genet Epidemiol. 34, 492-501.

Wellcome Trust Case Control Consortium. (2007). Genome-wide association study of 14,000 cases of seven common diseases and 3,000 shared controls. Nature. 447, 661-678.

Zeng D, Lin DY. (2005). Estimating haplotype-disease associations with pooled genotype data. . Genet Epidemiol , 28, 70-82.

Zuo Y, Zou G, Zhao H. (2006). Two-stage designs in case-control association analysis. Genetics. 173, 1747-1760.

Zuo Y, Zou G, Wang J, Zhao H, Liang H. (2008). Optimal two-stage design for case-control association analysis incorporating genotyp- ing errors. Ann Hum Genet. 72, 375-387.

Permissions

The contributors of this book come from diverse backgrounds, making this book a truly international effort. This book will bring forth new frontiers with its revolutionizing research information and detailed analysis of the nascent developments around the world.

We would like to thank Anjana Munshi, for lending her expertise to make the book truly unique. She has played a crucial role in the development of this book. Without her invaluable contribution this book wouldn't have been possible. She has made vital efforts to compile up to date information on the varied aspects of this subject to make this book a valuable addition to the collection of many professionals and students.

This book was conceptualized with the vision of imparting up-to-date information and advanced data in this field. To ensure the same, a matchless editorial board was set up. Every individual on the board went through rigorous rounds of assessment to prove their worth. After which they invested a large part of their time researching and compiling the most relevant data for our readers. Conferences and sessions were held from time to time between the editorial board and the contributing authors to present the data in the most comprehensible form. The editorial team has worked tirelessly to provide valuable and valid information to help people across the globe.

Every chapter published in this book has been scrutinized by our experts. Their significance has been extensively debated. The topics covered herein carry significant findings which will fuel the growth of the discipline. They may even be implemented as practical applications or may be referred to as a beginning point for another development. Chapters in this book were first published by InTech; hereby published with permission under the Creative Commons Attribution License or equivalent.

The editorial board has been involved in producing this book since its inception. They have spent rigorous hours researching and exploring the diverse topics which have resulted in the successful publishing of this book. They have passed on their knowledge of decades through this book. To expedite this challenging task, the publisher supported the team at every step. A small team of assistant editors was also appointed to further simplify the editing procedure and attain best results for the readers.

Our editorial team has been hand-picked from every corner of the world. Their multi-ethnicity adds dynamic inputs to the discussions which result in innovative outcomes. These outcomes are then further discussed with the researchers and contributors who give their valuable feedback and opinion regarding the same. The feedback is then collaborated with the researches and they are edited in a comprehensive manner to aid the understanding of the subject.

Apart from the editorial board, the designing team has also invested a significant amount of their time in understanding the subject and creating the most relevant covers. They scrutinized every image to scout for the most suitable representation of the subject and create an appropriate cover for the book.

The publishing team has been involved in this book since its early stages. They were actively engaged in every process, be it collecting the data, connecting with the contributors or procuring relevant information. The team has been an ardent support to the editorial, designing and production team. Their endless efforts to recruit the best for this project, has resulted in the accomplishment of this book. They are a veteran in the field of academics and their pool of knowledge is as vast as their experience in printing. Their expertise and guidance has proved useful at every step. Their uncompromising quality standards have made this book an exceptional effort. Their encouragement from time to time has been an inspiration for everyone.

The publisher and the editorial board hope that this book will prove to be a valuable piece of knowledge for researchers, students, practitioners and scholars across the globe.

List of Contributors

Sabrina Shore, Elena Hidalgo Ashrafi and Natasha Paul
TriLink BioTechnologies, Inc., USA

Bharti Rajendra Kumar
B.T. Kumaon Institute of Technology, Dwarahat, Almora, Uttarakhand, India

Rosemarie Tedeschi
University of Torino – DIVAPRA – Entomologia e Zoologia applicate all'Ambiente "C. Vidano", Italy

Laurent Granjon
Institut de Recherche Pour le Développement (IRD), CBGP (UMRIRD/INRA/CIRAD/MontpellierSupAgro), Campus de Bel-Air, Dakar, Senegal

Claudine Montgelard
Biogéographie et Ecologie des Vertébrés (EPHE), Centre d'Ecologie Fonctionnelle et Evolutive (UMR 5175 CNRS), Montpellier, Cedex 5, France

Victor Llaca
Dupont Agricultural Biotechnology, Pioneer Hi-Bred International, Wilmington, Delaware, USA

G. Darshan Raj
Department of Animal Husbandry and Veterinary Services, Government of Karnataka, India

Masayasu Kuwahara
Chemistry Laboratory of Artificial Biomolecules (CLAB), Graduate School of Engineering, Gunma University, Japan

Naoki Sugimoto
Frontier Institute for Biomolecular Engineering Research (FIBER) and Faculty of Frontiers of Innovative Research in Science and Technology (FIRST), Konan University, Japan

Chang-Yun Lin
McDermott Center of Human Growth and Development and Department of Clinical Sciences, University of Texas Southwestern Medical Center, Dallas, TX, USA

Tao Wang
Department of Epidemiology and Population Health, Albert Einstein College of Medicine, Bronx, NY, USA

Printed in the USA
CPSIA information can be obtained
at www.ICGtesting.com
JSHW011351221024
72173JS00003B/252